PHYSICAL PROCESSES IN LASERS

FIZICHESKIE PROTSESSY V OKG

ФИЗИЧЕСКИЕ ПРОЦЕССЫ В ОКГ

The Lebedev Physics Institute Series

Editor: Academician D. V. Skobel'tsyn

Director, P. N. Lebedev Physics Institute, Academy of Sciences of the USSR

Volume 25	Optical Methods of Investigating Solid Bodies
Volume 26	Cosmic Rays
Volume 27	Research in Molecular Spectroscopy
Volume 28	Radio Telescopes
Volume 29	Quantum Field Theory and Hydrodynamics
Volume 30	Physical Optics
Volume 31	Quantum Electronics in Lasers and Masers, Part 1
Volume 32	Plasma Physics
Volume 33	Studies of Nuclear Reactions
Volume 34	Photomesic and Photonuclear Processes
Volume 35	Electronic and Vibrational Spectra of Molecules
Volume 36	Photodisintegration of Nuclei in the Giant Resonance Region
Volume 37	Electrical and Optical Properties of Semiconductors
Volume 38	Wideband Cruciform Radio Telescope Research
Volume 39	Optical Studies in Liquids and Solids
Volume 40	Experimental Physics: Methods and Apparatus
Volume 41	The Nucleon Compton Effect at Low and Medium Energies
Volume 42	Electronics in Experimental Physics
Volume 43	Nonlinear Optics
Volume 44	Nuclear Physics and Interaction of Particles with Matter
Volume 45	Programming and Computer Techniques in Experimental Physics
Volume 46	Cosmic Rays and Nuclear Interactions at High Energies
Volume 47	Radio Astronomy: Instruments and Observations
Volume 48	Surface Properties of Semiconductors and Dynamics of Ionic Crystals
Volume 49	Quantum Electronics and Paramagnetic Resonance
Volume 50	Electroluminescence
Volume 51	Physics of Atomic Collisions
Volume 52	Quantum Electronics in Lasers and Masers, Part 2
Volume 53	Studies in Nuclear Physics
Volume 55	Optical Properties of Metals and Intermolecular Interactions
Volume 56	Physical Processes in Lasers

In preparation

Volume 54	Photomesic and Photonuclear Reactions and Investigation Methods with Synchrotrons
Volume 57	Theory of Interaction of Elementary Particles at High Energies
Volume 58	Investigations on Nonlinear Optics and Hyperacoustics
Volume 59	Luminescence and Nonlinear Optics
Volume 60	Spectroscopy of Laser Crystals
Volume 61	Theory of Plasmas
Volume 62	Stellar Atmospheres and Interplanetary Plasma: Techniques for Radioastronomical Devices
Volume 63	Nuclear Reactions and Interactions of Neutrons with Matter
Volume 64	Primary Cosmic Radiation
Volume 65	Stellarators
Volume 66	Theory of Coherent Acceleration of Particles and Emission of Relativistic Bunches
Volume 67	Physical Investigations of Strong Magnetic Fields
Volume 68	Emissive Recombination in Semiconducting Crystals
Volume 69	Nuclear Reactions and Accelerators of Charged Particles

Proceedings (Trudy) of the P. N. Lebedev Physics Institute

Volume 56

PHYSICAL PROCESSES IN LASERS

Edited by
Academician D. V. Skobel'tsyn
Director, P. N. Lebedev Physics Institute
Academy of Sciences of the USSR, Moscow

Translated from Russian by
James S. Wood

CONSULTANTS BUREAU
NEW YORK – LONDON
1973

The original Russian text was published by Nauka Press in Moscow in 1971 for the Academy of Sciences of the USSR as Volume 56 of the Proceedings of the P. N. Lebedev Institute. The present translation is published under an agreement with Mezhdunarodnaya Kniga, the Soviet book export agency.

Library of Congress Catalog Card Number 72-94826
ISBN 0-306-10884-4

A Division of Plenum Publishing Corporation
227 West 17th Street, New York, N. Y. 10011

United Kingdom edition published by Consultants Bureau, London
A Division of Plenum Publishing Company, Ltd.
Davis House (4th Floor), 8 Scrubs Lane, Harlesden, London, NW10 6SE, England

Printed in the United States of America

PREFACE

The present collection of articles sets forth the results of experimental research on physical processes in various types of optical quantum devices (lasers). Data are presented on the regulation of the temporal and spectral characteristics of solid-state lasers using a bleachable filter. A mode self-locking state is realized in which ultrashort light pulses are emitted. The stimulated emission in a pinch discharge due to transitions of singly and doubly ionized argon atoms in the visible and ultraviolet parts of the spectrum is investigated. Pulsed lasing associated with electron transitions in diatomic molecules at the leading edge of a powerful current pulse in a gas is subjected to systematic analysis.

The book is aimed at a broad spectrum of scientists and engineers specializing in quantum radiophysics.

CONTENTS

MODE DISCRIMINATION AND THE SELF-LOCKING EFFECT IN SOLID-STATE LASERS USING A BLEACHABLE FILTER

A. S. Markin

DETECTION AND INVESTIGATION OF STIMULATED EMISSION IN A PINCHED DISCHARGE

V. M. Sutovskii

PHYSICAL PROCESSES IN MOLECULAR HYDROGEN, DEUTERIUM, AND FIRST-POSITIVE-BAND-SYSTEM NITROGEN PULSED GAS-DISCHARGE LASERS

I. N. Knyazev

MODE DISCRIMINATION AND THE SELF-LOCKING EFFECT IN SOLID-STATE LASERS USING A BLEACHABLE FILTER*

A. S. Markin

INTRODUCTION

With the advent of the first lasers researchers devoted a great deal of effort, not only to searching for new types of active media and new mechanisms of population inversion, but also to developing new methods for controlling the emitted radiation characteristics of existing lasers. These characteristics include the laser output energy, peak power, directivity, spatial and temporal coherence, and the spectral composition of the radiation. One of these characteristics (or possibly several at once) can become preponderant, depending on the particular area of application of the laser (nonlinear optics, optical communications, holography, the production of a highly ionized plasma, etc.), and this characteristic (or set of characteristics) is the one that receives the greatest attention in a given type of laser. Our prime concern in the present study will be two such characteristics of laser emission, the peak power and spectral composition.

Generally recognized heretofore as one of the most effective methods for enhancing the peak power output of laser emission is the method of modulating the Q of the cavity ("Q switching" or "Q spoiling"). The method entails the following, in essence. Artificial losses are somehow induced in the laser cavity in such a way that the system will support self-sustained oscillations only at very high population inversion levels in the active medium. Then at the instant the population inversion attains its maximum value the artificial losses are abruptly (in the span of a few nanoseconds) removed. Inasmuch as the system gain is rather high at this moment, the de-excitation of atoms (or other excited particles) existing in negative-temperature states occurs very rapidly. This effect increases the peak power of the laser emission by several orders of magnitude relative to the power in the free-oscillation (unswitched) state.

Many methods are available at the present time for the rapid initiation and cessation of losses inside the cavity. In view of the great diversity of these methods it is convenient to think of them in several distinct groups. Among the principal groups are mechanical modula-

---

*Abridged text of the author's dissertation presented June 17, 1968, at the P. N. Lebedev Physics Institute of the Academy of Sciences of the USSR toward fulfillment of the academic degree of Candidate of Physicomathematical Sciences.

tion methods, such as the rotating disk [1] and rotating prism or mirror [2]; methods based on electro-optical effects involving a Kerr cell [3], Pockels cell [4], or Faraday cell [5]; and techniques utilizing the diffraction and refraction of light by ultrasonic waves [6, 7]. There are, in addition, many other Q-switching techniques that have not yet found widespread application, either because of their considerable complexity (such as the double Q-switching method [8, 9]) or due to their limited output characteristics (e.g., the exploding-film method [10], inhomogeneous magnetic field method [11], etc.). All of the foregoing techniques share one feature in common, namely the fact that the instant at which the shutter is opened is determined by an external signal, so that the emitted giant pulses are time-controlled.*

There is, however, a large class of shutters in which the opening time is not controlled by an external signal. For this reason, they have come to be known as passive shutters. The impossibility of exerting time control over the emitted pulses is the main shortcoming of this type of shutter, but it is superior to other types with respect to several other characteristics, and some of its attributes (emission spectrum selectivity and the mode-locking capability) are unique to this type of shutter. For the foregoing reasons, plus the consideration of extreme structural simplicity, passive shutters have lately come into widespread use for laser applications that do not require stringent time synchronization of the giant pulse.

Very recently another method of increasing the radiated power of laser emission has gained considerable recognition, namely the mode-locking method. The method comprises mainly the following. In most cases the phases of the different modes in a multimode laser (considering only axial modes) are independent, so that the beats between modes are smoothed out when there is a sufficiently large number of such modes. However, if the losses inside the cavity are modulated in some way at a frequency equal to (or a multiple of) the frequency difference between adjacent modes, definite relations are established between the phases of the individual modes. In the most elementary case the phase difference of the individual modes is a multiple of 2π. The laser emission in this case takes the form of a train of equally spaced pulses with a carrier frequency equal to the frequency difference between adjacent modes, $f_1 = c/2L$, where L is the optical path length of the cavity and c is the velocity of light. The width of these pulses is determined by the total number of locked modes and depends on the amplitude distribution of the latter. In the case of N_m locked modes having equal amplitudes the pulsewidth at the half-amplitude points is equal to $T_{pls} = 2.78 \cdot 2L/\pi c N_m \approx 1/c \cdot 2\Delta\nu_{em}$, where $2\Delta\nu_{em}$ is the width of the spectrum [12, 13]. The peak power of these pulses is N_m times the average power that occurs in the absence of Q switching and, hence, of mode locking.

Mode locking was first realized in gas lasers: neon–helium [14, 15] and argon [15], in which pulses having widths of ~0.5 and ~0.25 nsec, respectively, were generated. The pulse widths, as predicted by theory, turned out to be approximately equal to the reciprocal widths of the emission spectra of these lasers. It was also demonstrated in the cited investigations that in order to obtain good locking the modulation frequency and frequency difference between modes must be strictly equal, to within 10^{-4} [15]. This condition imposes severe demands on the stability of the modulation frequency in mode locking.

Somewhat later, mode locking was effected in solid-state lasers: ruby [16] and neodymium glass [17]; the effect was observed in both the free-oscillation state and in the giant pulse state. The single pulse widths in this case were ~1 nsec [16] and ~0.5 nsec [17]. It is important to note that solid state lasers hold more promise than gas lasers from the standpoint of peak power enhancement through the locking of all modes of the emission spectrum, because,

* The precision of this synchronization effect is limited by the instability of the time delay between opening of the shutter and the peak of the giant pulse due to, for example, the nonuniform distribution of the population inversion over the cross section of the active rod, or other factors.

first, even without locking their power is fairly high and can be further increased by Q switching, and, second, the width of the emission spectrum of the solid-state laser (particularly the neodymium glass variety) is considerably greater than the width of the stimulated emission spectrum of gas lasers, so that in the former it is possible to excite a far greater number of axial modes (with commensurate cavity lengths). However, another problem arises in connection with the locking of a large number of modes in the solid-state laser, namely, the maintenance of strict equality between the modulation frequency and intermodal frequency difference, because the latter can change during the emission process (as a result of, for example, heating of the active rod).

This problem is alleviated in some measure by what we call self mode locking, i.e., when mode locking is elicited by the nonlinear properties of some element of the laser (for example, saturation in a passive shutter). The self-locking effect was first observed in a gas laser [15] and subsequently in a neodymium glass laser (in our own work [18], as well as in [20]) and ruby laser [19] with a passive shutter ("passive Q switch"). The self-locking effect in a solid state laser with a passive shutter received the enthusiastic attention of many researchers for the added reason that its realization was expected to lead to the generation of ultra-short light pulses having widths on the order of 10^{-13} sec in the case of the neodymium glass laser [21-23] and 10^{-11} sec in the case of the ruby laser [23].

The anticipated potential of the self-locking method in lasers with a passive shutter with regard to the fabrication of sources of ultrashort very high-power radiation pulses imposed an urgent priority on the problem of finding methods for controlling the width and structure of the emission spectrum in this type of laser, because it is this characteristic of the emission which, in the final analysis, decides the ultimately obtainable pulse widths ($T_{pls} \approx 1/c \cdot 2\Delta\nu_{em}$). In addition, equal urgency was placed on the investigation of the nature of self mode locking as a function of various parameters of the laser for the purpose of realizing an operating regime in which all modes of the spectrum would be phase-synchronized. The solution of the stated problems called for concurrent investigations of the spectral and temporal characteristics of the emission of lasers incorporating passive shutters.

In view of the exceedingly broad sphere of problems whose solutions rely on the use of lasers with a passive shutter, heavier demands were placed on the temporal stability of the shutter parameters, and the need was recognized for far-reaching studies of the processes underlying the formation of giant pulses in the laser using a passive shutter. In most of the early studies carried out by means of lasers with a passive shutter, however, the required emission characteristics were attained by purely empirical means without sufficient analysis of the physics involved. Also, there is one other factor that underscores the timeliness of research on passive shutters. After some of the first papers had been published, dealing with passive shutters capable of meeting fairly strict requirements for a ruby laser [24-28], there appeared only one paper [29] in which a passive shutter for a neodymium glass laser was described. And, as our investigations had already demonstrated, the solution used as the shutter for the given laser failed to meet practical demands on account of its low thermal and photochemical stability, which tended to hasten its decomposition.

Nevertheless, the neodymium glass laser, like the ruby type, remains today as one of the most powerful solid-state quantum oscillators in existence. Moreover, the considerable width of the luminescence line of neodymium glass (by comparison with ruby) lends sizable advantages to this type of laser for the solution of the problem of obtaining ultrashort pulses. For these reasons, the creation of a passive shutter for neodymium glass lasers that will meet the ever-growing requirements has become one of the most pressing practical problems.

Another significant characteristic, besides the peak power of laser emission, is its spectral composition. The demands imposed on this characteristic can differ widely, depending on the area of application of the laser. Thus, for several branches of nonlinear optics (various types of induced scattering and parametric oscillators), holography, and other applications, coherent radiation with an extremely high degree of monochromaticity is required, i.e., a laser is wanted that will operate in exactly one axial mode. For other problems (for instance, the generation of ultrashort light pulses on the basis of mode locking) a laser is wanted that has the broadest possible emission spectrum. Moreover, the actual problem of controlling the emission spectrum of a laser is one of independent interest in that it is related to the mechanism of the formation of the radiation pulse. The parameters and processes that can be controlled in the given case are various elements of the cavity, as well as the emission buildup regime. It is essential to note that the detailed investigation of the influence of the cavity elements on the emission spectrum of lasers will make it possible to solve the converse problem, i.e., to draw certain conclusions with regard to the laser emission buildup regime on the basis of a study of the variations elicited in the spectrum by the selective properties of a particular element.

The fundamental problem set forth in the present study was to conduct systematic investigations of the principal (temporal and spectral) properties of lasers using a passive shutter so as to exhibit the specifics of its operation and to develop techniques for controlling the emission characteristics of such lasers in the required direction. Another, narrower practical objective was to study solutions of a number of new substances for the purpose of obtaining a passive shutter for neodymium glass lasers that would have higher operational characteristics than those currently described in the literature [29].

The paper is divided into four chapters. Chapter I is concerned with the realization and investigation of Q switching in a neodymium glass laser by means of a liquid bleachable filter. Physical justification is given for the requirements imposed on the parameters of the substances designed to function as passive shutters. Also described in the same chapter are the investigational procedure and schematic of the passive-shutter laser used in the study. The effect of the emission regime on the temporal characteristics of the resulting giant pulses is investigated. The experimental results are discussed on the basis of the giant pulse theory developed earlier for an instantaneously actuated shutter ("ideal switch").

The experimental results of an investigation of the discrimination of axial modes in a neodymium glass laser with Q switching by a passive shutter and in the free-oscillation (unswitched) regime are presented in Chapter II. The investigation has been carried out both by the direct spectroscopic method and by the beat-frequency measurement technique. The results of the observations are compared with the calculated characteristics obtained from a computation of the Q factors of different modes in a complex cavity. The influence of various elements of the cavity and lasing regime on the emission spectrum is discussed and analyzed. Attention is focused on certain new laws and effects never before observed. For the purpose of exhibiting the peculiarities of the behavior of the lasers in different operating regimes some of the investigations have been conducted in the Q-switching regime using a rotating prism. Moreover, control experiments have also been conducted with a ruby laser having a passive shutter in order to generalize the results on mode discrimination by various cavity elements in the Q-switching regime.

Chapter III is devoted to an investigation of the mechanism of radiation pulse formation in the Q-switching regime with the use of a passive shutter and in the free-oscillation regime. The new method of investigation used here relies on earlier results on mode discrimination. The proposed method is used to estimate the temporal development of a giant pulse in a neodymium glass laser with a passive shutter, as well as the buildup time of the individual free-oscillation spike. The results are discussed on the basis of a simplified model of the development of a giant pulse and free-oscillation spike. The data of these investigations make it possible to formulate the mechanism of the buildup of the emission pulse in both regimes and to estimate the discrimination order that must be introduced in the cavity for the suppression of unwanted modes.

The results of an investigation of the self-locking effect in neodymium glass and ruby lasers using a passive shutter are given in Chapter IV. The influence of the position of the cell containing the bleachable solution within the cavity on the nature of the self-locking effect is analyzed. The results are discussed on the basis of a theoretical model of self mode locking in a laser using a bleachable filter. The ultimate width of the ultrashort pulses in specific laser arrangements with a passive shutter is discussed on the basis of experimental results from a measurement of the intensity distribution in the emission spectra in the self-locking regime and a simplified model of pulse buildup in this type of laser.

CHAPTER I

Q SWITCHING OF LASERS BY MEANS OF A PASSIVE SHUTTER

§1. Operating Principle and Types of Passive Shutters

As stated in the Introduction, the Q switching of lasers can be realized by means of a passive shutter. The operating principle of this type of shutter entails the following. A passive element, whose transmissivity at the emission wavelength depends on the power of the light flux incident upon it, is placed in the laser cavity which is formed by external mirrors and is only partially filled with the active medium. The insertion of this element into the cavity increases the initial losses and, hence, the pumping threshold level. As a result, the population inversion at the initial instant of emission will be considerably higher than in the absence of the passive element. The onset of emission inside the cavity (prior to the instant of "bleaching" of the passive element this is the so-called free-oscillation regime) causes a rapid increase in the light flux density through the element, so that when this density attains a certain threshold level the transmissivity of the element drops to an insignificant value, i.e., "bleaching" of the passive element takes place. Inasmuch as the oscillation buildup time is normally small (relative to the pumping time, spontaneous decay time of the upper level, etc.) and only a minute portion of the population inversion is spent in bleaching of the passive element, the entire remaining store of particles after bleaching undergoes de-excitation in the form of a powerful short light pulse, i.e., a giant pulse.

If a particular material is to be used for the passive element, it must meet a number of requirements. It must, above all, have an absorption band in the vicinity of the laser emission wavelength. Moreover, such characteristics of the passive element as the cross section σ_p for induced transition at the emission wavelength and the decay time t_p of the upper level govern the minimum flight flux Φ_{thr} that is required in the cavity for bleaching of the element. The relationship between these variables has the most elementary form in the case of a passive element having an idealized two-level scheme. In this case the kinetic equation for the difference between the population densities M_1 and M_2 at the lower and upper levels is written in the form

$$\frac{dM}{dt} = -2\sigma_p M\Phi + \frac{M_0 - M}{t_p}, \tag{1}$$

where $M = M_1 - M_2$, $M_0 = M_1 + M_2$, and Φ is the light flux density (photons/cm^2 sec). In the steady state $dM/dt = 0$, and the following expression obtains for the transmissivity of the given system:

$$T(\Phi) = \exp(-M\sigma_p l_p) = \exp\left(-\frac{M_0\sigma_p l_p}{1+\Phi/\Phi_{thr}}\right), \tag{2}$$

in which l_p is the length of the passive element and $\Phi_{thr} = 1/2\sigma_p t_p$ is conditionally referred to as the threshold light flux density, above which the transmissivity drops sharply. For fluxes less than Φ_{thr}, on the other hand, the transmissivity of the system is close to the initial value $T_0 = \exp(-M_0\sigma_p l_p)$, i.e., to the transmissivity at zero flux. We give as an example a solution of phthalocyanine in nitrobenzene [30]. In this case $\sigma_p \approx 0.3 \cdot 10^{-15}$ cm^2, and $t_p \approx 3$ nsec, so that for Φ_{thr} we obtain $\sim 1.6 \cdot 10^5$ W/cm^2. This value is consistent with the experimental results [31].

A concurrent analysis of the set of kinetic equations describing the behavior of the active and passive particle densities, as well as the photon density in the cavity (such an analysis was first carried out in [32, 33]) shows that emission in this kind of system is possible only if the following inequality holds:

$$\sigma = \frac{\sigma_p}{\sigma_a} > \frac{n'_{ai}}{n'_{ai} - 1}, \tag{3}$$

in which σ_a is the induced transition cross section for active particles and $n'_{ai} = n_{ai}/n_p$ is the population inversion coefficient, which is equal to the ratio of the initial population inversion density of active particles, $n_{ai} = (N_2 - N_1)/N_0 = (N_2 - N_1)/(N_1 + N_2)$, to the threshold inverted population density in the absence of the passive element. According to this condition the generation of a giant pulse requires a passive element for which, at least, $\sigma_p > \sigma_a$.

There are now several types of passive shutters (passive switches) in existence. The most plentiful and widely used class of elements of this type includes liquid solutions of certain organic dyes [24-29, 34-39]. Certain kinds of glass light filters [40, 41], in which the transmission limit is shifted under the action of powerful light radiation; dye films deposited on a glass substrate [10, 42]; and polaroid films [43] have been used with measurable success as passive shutters. Passive Q switching has also been effected by dispersing particles of an absorber among the active particles in glasses containing both active ions and absorption centers as an impurity [44-46]. Very recently certain gases and gas mixtures have been successfully used as passive shutters [47, 48].

§2. Q Switching of Neodymium Glass Lasers by Means of a Bleachable Liquid Filter

At the time the present study was begun the situation was as follows with regard to bleachable-filter Q switches for lasers. Several types of bleachable filters had been used with success for ruby lasers: liquid solutions of phthalocyanine [24, 27], cryptocyanine [25, 26], and certain types of glass light filters [40, 41]. With these bleachable filters it was possible to generate giant pulses with a power of tens of megawatts and widths up to 10 nsec, and the characteristics of the given shutters remained unaltered after 500 or more pulses. At that time only one substance was known in the literature as a passive shutter for neodymium glass lasers, namely polymethine dye [29], which in solution with methyl alcohol had made it possible to obtain a giant pulse with a power of ~1 MW and a width of ~25 nsec. However, as our investigations revealed, the properties of this solution were rather quickly altered (toward an increase in the initial transmissivity) due to photochemical and thermal decomposition, an appreciable increase in the transmissivity occurring both in interaction with powerful laser radiation (photochemical decomposition after several bursts at a power of ~50 MW/cm^2) and simply with the passage of time (thermal decomposition in a matter of days).

Our first problem, therefore, was to try to find new substances* whose solutions could be used as passive shutters for neodymium glass lasers, as well as to investigate and compare their properties so as to learn which had the most stable characteristics with respect to time wear and radiation power. It is important to note that the stability of the passive shutter material is important not only from the standpoint of the practical application of lasers and the generation of large powers, but also with regard to research on the physical processes involved in this kind of laser, in which operational stability is significant.

The laser cavity used in these experiments is illustrated schematically in Fig. 1. The cavity was formed by two mirrors R_1 and R_2 comprising dielectric coatings deposited on plane-parallel glass substrates. The reflectivity of the first mirror at the emission wavelength (λ = 1.06 μ) was made as large as possible ($R_1 \gtrsim 0.99$), and the reflectivity of the second was chosen between the limits R_2 0.15 to 0.70, depending on the pumping level, where the values of R_2 = 0.15 to 0.30 were usually realized by using a glass substrate with different values of the refractive index n and no reflective dielectric coatings. The maximum reflectivity of this substrate, with interference effects taken into account, is equal to the following, according to [49]:

$$R_{max} = \left(\frac{2r}{1+r^2}\right)^2, \tag{4}$$

where r = (n − 1)/(n + 1) is the Fresnel amplitude reflection coefficient at the air–glass interface.

Inside the cavity was a light source L with active rod AR and cell K containing the dye solution to be used as the passive shutter. The active rods were type KGSS-7 neodymium glass rods 8 to 15 mm in diameter and of length l_a ~ 120 mm. The ends of the rods were plane parallel to within a maximum error of 10″. The lateral surface of the rods was either matte or etched so as to preclude the generation of nonaxial modes [50, 51]. Optical pumping of the rod was realized by means of two IFP-2000 pulsed lamps PL connected in series and supplied by a bank of capacitors having total capacitance C = 1000 μF. The voltage on the capacitors could be varied from 700 V to 3 kV, so that the range of variation of the electrical pumping energy was from ~250 J to 4.5 kJ. A typical time curve of the pumping intensity is shown in Fig. 2. As the figure indicates, the pumping period at the half-peak points was about 400 μsec. The pumping lamps and active rod were placed together in a figure-eight elliptical bulb, and their attachment was such that the lamps and rod could be cooled with circulating water. The inner

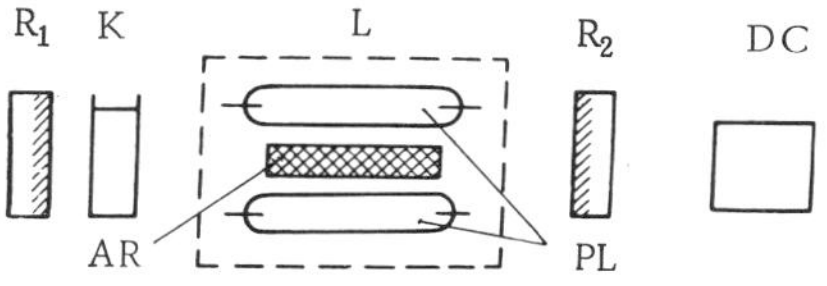

Fig. 1. Experimental arrangements. R_1, R_2) Cavity mirrors; L) light source; K) cell; AR) active rod; PL) pumping lamps; DC) detection circuit.

*Some phototropic substances were synthesized at the Scientific-Research Institute of the Chemistry of Photographic Materials (NIIKhIMFOTO) under the direction of I. I. Levkoev and A. F. Vompe, to whom the author conveys his sincere appreciation.

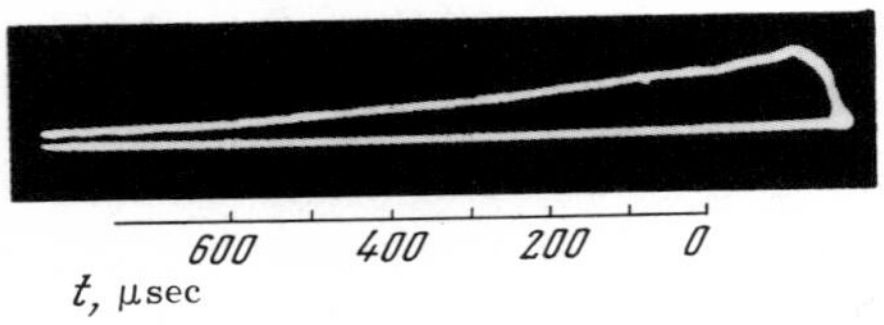

Fig. 2. Oscillogram of pumping intensity (recorded on an S1-16 oscilloscope and FÉU-17 photoelectric detector).

surface of the bulb was silvered and polished to increase the pumping efficiency. The elliptical cross section of the bulb, of course, prevented the cross section of the active rod from becoming uniformly filled with the pumping radiation, but this type of light source had a fairly high efficiency and, in the case of neodymium glass rods 10 mm in diameter and 120 mm long, ensured a population inversion such that single-pass gain of seven was attained for light transmitted through the rod.

In our work with liquid solutions we used two constructions for the cell K. The first construction (Fig. 3a) comprised an intermediate ring of thickness d_r = 3 to 10 mm with windows on both sides in the direction of optical contact. The parallelism error of all surfaces in the assembly did not exceed 10". In order to reduce the air contact of the solution the inlet opening was closed off with a ground glass plate. The inside diameter of the intermediate ring was calculated to be approximately equal to the diameter of the light beam. With this cell construction the entire volume of the solution interacted with the radiation, i.e., the entire volume of the cell was the working volume. This cell construction was used to investigate the comparative stability of various types of solutions used as passive shutters. The active rod in these studies was 20 mm in diameter and 240 mm long, and a total-internal-reflecting prism (R_1) and plane-parallel glass substrate with $n \approx 1.7$ (R_2) were used for the cavity mirrors. The use of these elements in the cavity made it possible at a radiant power density of ~100 MW/cm^2 to generate up to 500 bursts or more without any change in the system parameters (excepting the passive shutter), a feature that is particularly important for comparative studies.

The second cell construction (Fig. 3b) was provided with two solution-filled volumes: the working volume V_w, in which the solution interacted with the radiation; and the ballast volume V_b, in which the radiation did not act on the solution. This construction makes it possible to increase the operating life of the solution in the cell as a whole when decomposition of the substances takes place mainly as a result of interaction with the laser radiation. Experience showed that this type of cell with a thickness d_r = 3 mm could be used with certain solutions to generate up to 100 bursts at a power of ~100 MW/cm^2 without any appreciable change in the

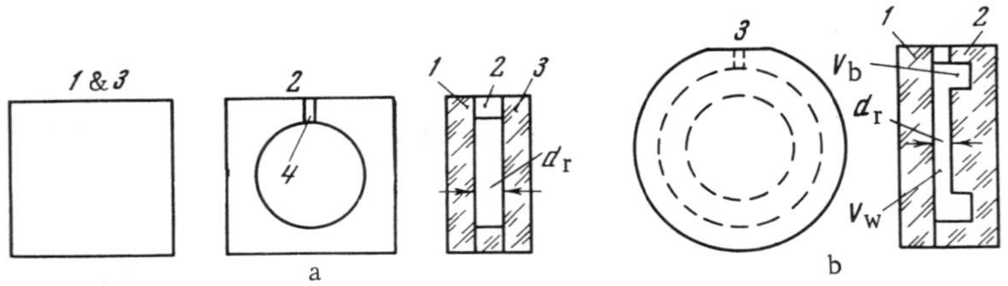

Fig. 3. Constructions of cells for the liquid passive shutter. a) Construction without ballast volume; 1, 3) cell windows 5 to 10 mm thick; 2) base of cell; 4) opening for solution input. b) Construction with ballast volume V_b: 1) cell window; 2) cell base; 3) inlet opening.

shutter characteristics, whereas with the first type of cell, given the same thickness, after the same number of bursts (at the same power) the transmissivity of the cell containing the same solution increased from $T_0 = 0.30$ to 0.35 or 0.40. Also, the second cell construction has several other qualities in its favor. First, it has only one optical contact, so that its strength is greater, and, second, the thickness of the cell can be made considerably smaller than that of the first type, a result that is especially important in connection with the use of solvents having a large absorption at the emission wavelength.

The transmissivities of the cells containing the bleachable solutions were measured with a Hitachi type EPS-2U double-beam spectrophotometer. The spectral width of the slit in these measurements did not exceed 30 Å at $\lambda = 1.06\,\mu$. For more precise measurements a domestic type SF-8 double-beam spectrophotometer with a resolution of ~2.5 Å in this wavelength range was used.

The emitted giant pulses were detected both with an IMO-1 calorimeter and with a type FÉK high-speed photocell. Inasmuch as the range of linear currents for this type of photocell extends to several amperes, with a load $R_L = 75\,\Omega$ this signal normally sufficed for operation without an amplifier, i.e., the electrical signal from the photocell was sent directly to the cathode-ray tube of the oscilloscope. As our subsequent experiments showed, the time constant of the FÉK-09 plus the cathode-ray tube of the S1-11 oscilloscope was on the order of 2.0 to 2.5 nsec, the time resolution being limited in this case by the characteristics of the tube. The time constant of the detection system could be decreased to 0.8 to 1.0 nsec by using a type I2-7 oscilloscope.

In our first experiments the passive element was a cell containing a solution of polymethine dye in methyl alcohol; the dye had the same composition as in [29]. Under certain pumping conditions, i.e., in an operating state such that the pumping energy only slightly (≤10%) exceeds the energy threshold for self-excitation of the laser and cell, the emission from the laser has the form of a single pulse whose energy and width depend on the initial transmissivity of the solution. The time variations of the luminescence of a KGSS-7 active rod and the emission of a neodymium glass laser with Q switching by means of a bleachable filter are shown in Fig. 4a and in Figs. 4b and 4c, respectively. For near-threshold pumping (Fig. 4b) the emission is a single giant pulse situated on the time scale very close to the peak luminescence intensity. When the pumping energy considerably exceeds the threshold value (Fig. 4c), several giant pulses are emitted, the first situated to the left of the luminescence peak and the

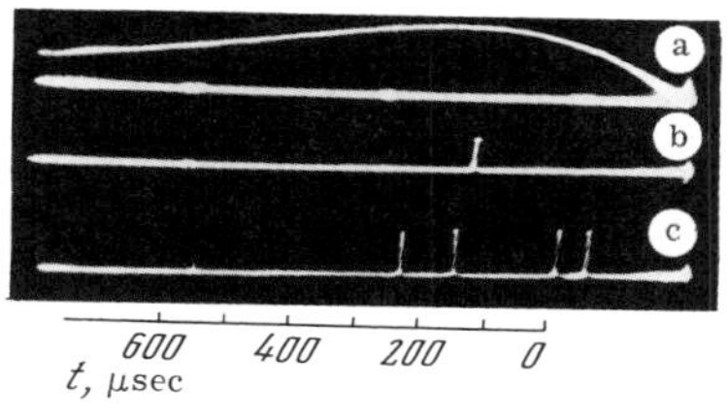

Fig. 4. Oscillograms of the emitted luminescence of KGSS-7 neodymium glass (a) and the emission of a neodymium glass laser with passive Q switching (b, c). b) Pumping energy at most 10% above threshold; c) pumping energy three times the threshold value.

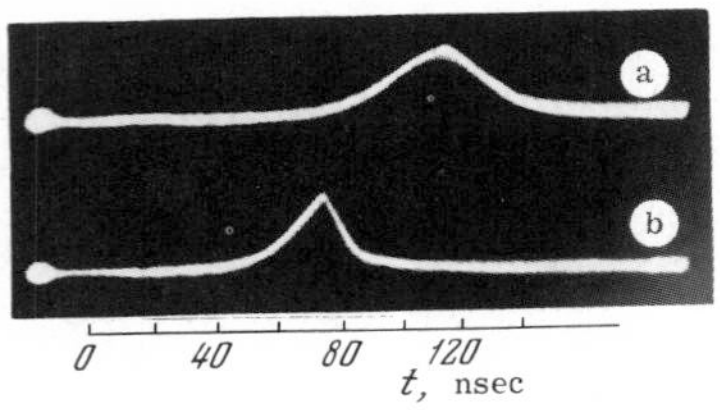

Fig. 5. Oscillograms of giant pulse emission after the focus of a lens $f = 50$ mm. a) No spark formation at focus; b) spark formation.

last to its right. This means that the population inversion required for emission was reached before the maximum population inversion was attained.

It was demonstrated even in our first experiments [35] with a 12-mm-diameter, 120-mm long KGSS-7 sample that a single giant pulse with a power of ~5 MW and width of ~35 nsec (Fig. 5a) could be obtained using a cell filled with a solution of polymethine dye of the type used in [29] in methyl alcohol ($T_0 = 40\%$). When this emission was focused by a lens with $f =$ 50 mm, air breakdown in the form of a spark was observed at the focus of the lens. Analyzing the time dependence of the laser emission after the lens focus, we noted that in the presence of a spark the shape of the transmitted pulse was sharply altered (Fig. 5b). This shortening of the trailing edge of the transmitted pulse, an effect that was observed simultaneously and independently with a ruby laser in [52] and was explained in this paper, was subsequently used as a method for the shaping of light pulses with a steep trailing edge [53].

Using a polymethine dye of the type used in [29] (or, in our classification [54], No. 6), we observed a rather fast increase in the transmissivity of its solutions either when exposed to light or in interaction with powerful laser emission. This increase was irreversible and was elicited by the photochemical decomposition of the dye. We undertook investigations of new substances and solvents from the point of view of selecting the most stable combinations. With a solution of one such dye (No. 3 in the system of [54]) in nitrobenzene in a laser containing a KGSS-7 active rod 12 mm in diameter and 120 mm long we were able to produce a giant pulse with a peak power of ~50 MW [55]. This power was sufficient for the formation of a spark at the focus of a lens $f = 500$ mm. The width of the generated pulse was ~10 nsec (Fig. 6) and for the selected parameters of the system turned out to be quite close to the value predicted by the theory of a shutter with instantaneous switching [56], i.e., the ultimately attainable value by this Q-switching technique. The system parameters were as follows: The active rod with an etched lateral surface had the above-indicated dimensions, the reflectivities of the mirrors were equal to $R_1 \gtrsim 0.99$ and $R_2 \approx 0.40$, the initial transmissivity of the cell containing the bleachable solution at the emission wavelength $\lambda = 1.06\,\mu$ was $T_0 \approx 0.20$, and the optical path-length of the cavity was $L = 55$ cm.

Using a 16-mm-diameter, 240-mm-long KGSS-7 rod as the active medium ($R_1 \gtrsim 0.99$, $R_2 = 0.3$ to 0.4, $T_0 = 0.1$ to 0.2), we obtained a single pulse with a peak power up to 250 MW and width of 10 to 15 nsec. It is essential to note in this connection that the continued power growth

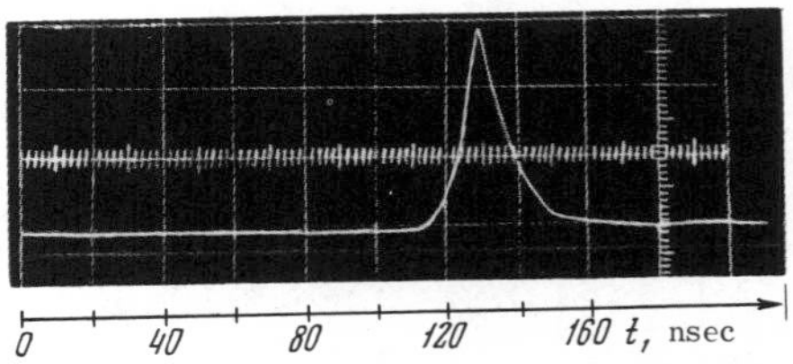

Fig. 6. Oscillogram of a giant pulse from a neodymium glass laser with a passive shutter (No. 3 dye in nitrobenzene).

(with a reduction in T_0 and increased pumping) was limited by the destruction of the active rods at the time of the first burst.

Consequently, we were able for the first time in the case of a neodymium glass laser to realize a liquid passive shutter which in a certain sense could be regarded as "ideal," since the width of the giant pulse from a laser using this shutter agrees with the result predicted by the theory of an instantaneous switch [56]. The future perfection of this type of shutter therefore had to be aimed at improvement of the stability of its characteristics. The attempt to increase the peak power of the laser by continuing to decrease the width of the giant pulses had to be tied in with searches for new mechanisms of Q regulation.

§3. Dependence of the Width and Shape of the Giant Pulse on the Population Inversion Coefficient

When passive Q switches first made their appearance, several theoretical papers had been published on the formation of giant pulses in a Q-switched laser [8, 56-59]; in these papers the dependence of the characteristics of the giant pulse on various parameters of the laser was investigated. Passive Q switching was not discussed in these papers, naturally, and the switch was postulated either as instantaneous [56, 57] or as obeying a certain law [8, 58, 59]. There had been no experimental studies conducted with the object of quantitatively checking out the theoretical results at that time. Given definite assumptions, however, with regard to the mechanism of the buildup of the giant pulse in a laser with a passive shutter, some of the results of these theoretical studies could be compared quantitatively with the observed characteristics of the giant pulse produced by a laser with a passive shutter.

We decided on the work of Wagner and Lenguel [56] as the initial theoretical study whose results were to be experimentally tested. These authors analyzed the two-level model of a laser in single-mode operation and neglected effects whose rates were slower than the giant pulse width (spontaneous emission, variation of the upper level population due to pumping, etc.). The Q switching of the cavity was assumed to be instantaneous. Proceeding on the basis of these assumptions, the authors of [56] derived rate equations for the photon density Φ in the cavity and the inverted population density $N = N_2 - N_1$ as follows:

$$\frac{d\Phi}{dt} = \left(\frac{\alpha l_a}{t_c} - \frac{1}{t_{ph}}\right)\Phi, \quad \frac{dN}{dt} = -\frac{2\alpha l_a}{t_c}\Phi, \tag{5}$$

in which l_a is the length of the active rod of the laser, $\alpha = \alpha_0 N/N_0$ is the gain (α_0 is the absorption coefficient of the active medium in a working transition without excitation; $N_0 = N_1 + N_2$), $t_c = 2L/c$ is the round-trip transit time of a photon over the optical path of the cavity, and t_{ph} is the photon lifetime in the cavity. In turn, $t_{ph} = t_c/\varkappa$, where $\varkappa = \varkappa_1 + \varkappa_2$ is the total loss factor (i.e., the fraction of annihilated photons relative to their total number in time t_c), and $\varkappa_1$ and $\varkappa_2$ are the loss factors due to emission of radiation from the cavity and parasitic losses inside the cavity.

The solution of the set of equations (5) yields analytic expressions for the total giant pulse energy E and peak power W_P as a function of the parameters N_i, N_P, N_f, and V, where N_i is the initial inverted population density at the instant of Q switching, N_P is the population inversion density at the instant the pulse attains peak power, N_f is the final inverted population density, and V is the volume of the active rod. A visual representation of these parameters and a qualitative picture of the variation of N and Φ are given in Fig. 7. Wagner and Lenguel also used a digital computer to calculate the durations t_1 and t_2 of the leading and trailing edges of the pulse for given ratios of the population inversion coefficient $n'_{ai} = N_i/N_P$. They

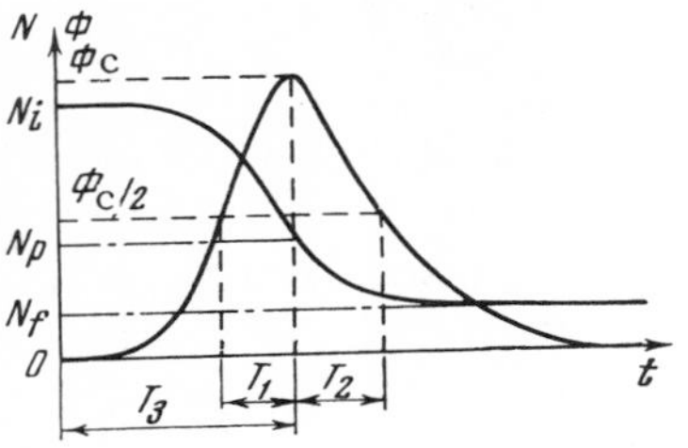

Fig. 7. Time variation of population inversion density N and photon density Φ in the cavity in the case of an ideal shutter (according to [56]).

presented the results of their calculation in the form of a table, which gave the numerical values of the quantities $\tau_1 = t_1/t_{ph}$ and $\tau_2 = t_2/t_{ph}$ for different values of n'_{ai} within the limits from 1.1 to 12.0. Their results show, in particular, that as the ratio n'_{ai} is increased the total pulse width $t_1 + t_2$ decreases, while the pulse itself becomes skewed toward a steeper leading edge.

A comparison between theory and experiment is possible in principle by direct measurement of the quantities E, W_P, $\tau_1 + \tau_2$, and τ_2/τ_1 and comparison of their values with those calculated in [56] on the basis of measurements of N_i, N_P, and V. Prior to the publication of our paper [60], however, a quantitative comparison of this nature had not been made. The main reason for this was contained in the patent difficulties associated with measurement of N_i, N_P, and the actual effective volume V of the active rod. Moreover, the majority of Q switches existing at that time (rotating prisms, Kerr cells, Pockels cells, etc.) had the attribute that they "switched on" the entire region of the active rod cross section in which the required population inversion $N_i > N_P$ was attained. It is well known, however, that the population inversion density distribution, for a number of reasons (optical inhomogeneity of the rods, inhomogeneity of the optical pumping, onset of oscillations in internal modes, etc.), is not uniform over the rod cross section. As a result, different parts of the rod cross sections participate in the formation of the giant pulse with different values of N_i, so that the latter quantity becomes indeterminate. The actual observed pulse is therefore averaged over the entire rod cross section, thus complicating the comparison of theory with experiment.

Passive shutters have a well-known advantage from the standpoint of comparison between theory and experiment. When Q switches of this type are used, the operating regime of the laser is such that the actuation of the shutter (bleaching of the filter) takes place as a result of self-excitation of the system. In this case, when the pumping energy slightly exceeds the threshold value (bearing in mind the threshold of self-excitation of the laser and cell), oscillation occurs only in those regions of the rod cross section in which the gain exceeds the total losses (losses at the mirrors, in the active medium, and in the passive element) when the required population inversion density is attained. Under such conditions only definite regions of the rod cross section with maximum N_i will take part in the formation of the giant pulse. This is why, clearly, the least attainable giant pulse width using earlier types of shutters (rotating prisms, Kerr cells, etc.) was 20 or 30 nsec. With passive shutters, on the other hand, pulse widths on the order of 10 nsec were obtained with comparable pumping powers and similar active rods.

The existence of well-defined regions of the rod cross section that participate in the formation of the giant pulse is experimentally confirmed by photographing the radiant energy distribution in the near zone. The operating regime of lasers with a bleachable liquid filter in this case was controlled in such a way as to investigate a single pulse. A typical radiant energy distribution over the end face of a neodymium glass rod is illustrated in Fig. 8, in which the confined regions taking part in the emission process are clearly visible. It is very difficult in practice, however, to determine the rod volume that actually participates in the forma-

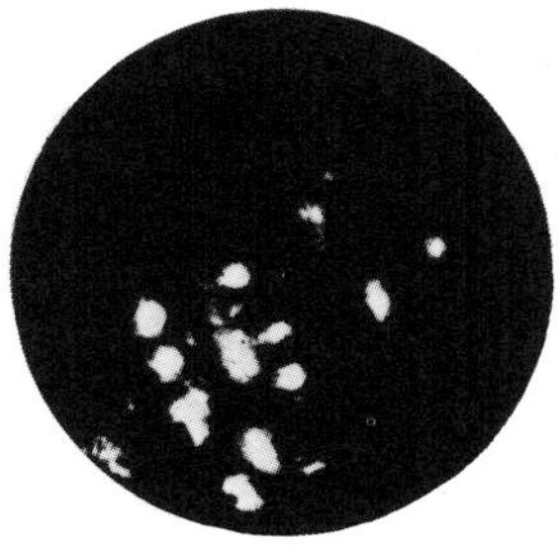

Fig. 8. Radiant energy distribution of a giant pulse in the near zone ($T_0 = 0.7$; rod diameter, 11 mm).

tion of the pulse, and it is therefore difficult to draw comparisons between theory and experiment for quantities involving the volume V, i.e., for E and W_P.* On the other hand, it is relatively easy in the case of passive Q switching to determine the population inversion coefficient n'_{ai} and to compare the theoretical and experimental values of the quantities $\tau_1 + \tau_2$ and τ_2/τ_1.

The method proposed in our work for experimentally determining the values of n'_{ai} is based on the following relations. For a passive shutter, assuming that the main losses in the cavity are solely attributable to absorption losses in the shutter and losses due to the emission of radiation from the cavity through the mirrors, the emission threshold condition may be written in the form

$$R_1 R_2 T_0^2 \exp(2\sigma_a N_i l_a) = 1, \tag{6}$$

where R_1 and R_2 are the reflectivities of the cavity mirrors, T_0 is the initial transmissivity of the filter (in the unbleached state) at the emission frequency, and σ_a is the induced-transition cross section for particles of the active medium at the emission frequency.

At the instant that the giant pulse attains peak power (assuming that the bleaching of the filter is complete and sufficiently rapid) the total losses are equal to the gain in the active medium with population inversion density N_P, so that at the given instant the following relation must hold:

$$R_1 R_2 \exp(2\sigma_a N_P l_a) = 1. \tag{7}$$

An expression for the population inversion coefficient n'_{ai} as a function of simple parameters is easily obtained from relations (6) and (7):

$$n'_{ai} = \frac{N_i}{N_P} = \frac{\ln(R_1 R_2 T_0^2)}{\ln(R_1 R_2)}. \tag{8}$$

Thus, by varying the value of the initial transmissivity T_0 of the passive shutter it is possible to exercise a controlled variation of the coefficient n'_{ai} for fixed values of R_1 and R_2. Measuring the pulse width $t_1 + t_2$ and the ratio t_2/t_1 for each value of n'_{ai}, one can compare the results directly with the results of the theoretical calculations [56].

The experimental arrangement was the same as described in §2. The experiment was carried out with cylindrical KGSS-7 rods of various dimensions. The bleachable filter in this case was a solution of No. 7 dye (see [54]) in nitrobenzene. The ends of the neodymium glass rod and planes of the cell containing the bleachable solution were normally inclined at a certain

* It is possible in this case, however, to pose the converse problem, i.e., to determine the effective rod volume V_{eff} contributing to emission by measuring the total pulse energy E and initial population inversion N_i.

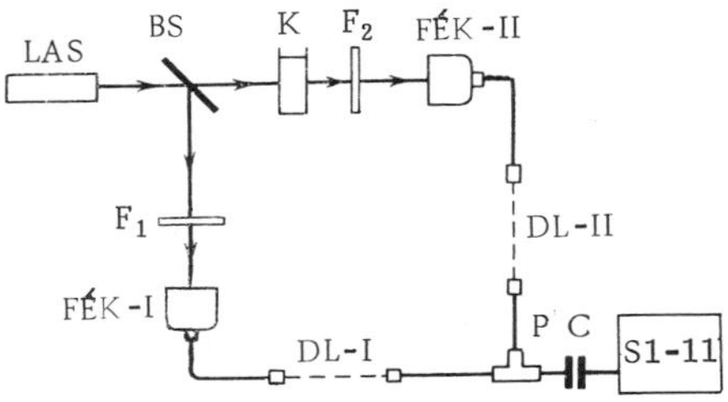

Fig. 9. Experimental arrangement for determining the transmissivity of the bleachable filter with powerful light fluxes. LAS) Neodymium glass laser with passive shutter, power 30-50 MW/cm^2; BS) beam-splitting plate; K) test cell; F_1, F_2) calibrated neutral filters; FÉK-I, FÉK-II) coaxial photocells; DL-I) 35-nsec delay line; DL-II) 220-nsec delay line; D) two-way divider; C) separative capacitor.

angle (~ 30') relative to the planes of the mirrors during adjustment. The transmissivity was varied from 0.35 to 0.98 by varying the dye concentration. The value of T_0 was measured with a maximum absolute error of 0.005. The value of R_2 was made equal to 0.5 to 0.6. The time resolution of the detection system (FÉK-09 photocell plus cathode-ray tube of S1-11 oscilloscope) was about $(2.0 \text{ to } 2.5) \cdot 10^{-9}$ sec.

Inasmuch as the induced-transition cross section for particles of the passive shutter is usually very large ($\sigma_p = 10^{-15}$ to 10^{-16} cm^2) relative to the cross section for active particles ($\sigma_a \approx 10^{-20}$ cm^2), for giant pulse widths even one or two orders of magnitude greater than the decay time of the upper level for particles of the passive shutter only an insignificant fraction of the total pulse energy must be expended in bleaching of the filter. Consequently, the quantity N_i in relation (6) will be very close to the true initial population inversion involved in the equations of [56]. To test the validity of this assumption in our case we conducted the following experiment (Fig. 9). The radiation from a laser with a passive shutter (the transmissivity of the solution in the cavity was equal to $T_0 \approx 0.4$) was split by a glass plate into two beams, which were detected independently by two detectors (FÉK-09 photocell plus CRT of S1-11 oscilloscope). The beam intensities were equalized by neutral filters so that both signals had the same amplitude on the oscilloscope. Then a cell containing a solution of the same dye as in the laser (No. 7 in nitrobenzene) was inserted into the beam with the greater intensity (transmitted through the glass beam splitter). The measurements showed that, within the experimental error limits (~ 20%), the intensity variations of the beam transmitted through the cell with the solution do not occur, down to very small ($T_0 \sim 0.001$) values of the transmissivity of the investigated cell (Fig. 10). Consequently, it was reasonable to assume in our case that a very small fraction of the pulse energy was spent in bleaching of the filter used in the experiments (for large enough powers) and that the bleaching of the filter was complete.*

In order to obtain more reliable data on the width of the giant pulses, for each value of the transmissivity T_0 we photographed several (five to seven) pulses and averaged the values obtained for t_1 and t_2. In the averaging we disregarded pulses having a complex waveform (which were comprised in 40% of the total number), i.e., pulses having two or more peaks. Such pulses, it seemed to us, were the possible result of mode interaction, and their shape could not be described by the theory of [56], which was formulated for single-mode lasers.

* Complete bleaching of the passive filter (phthalocyanine solutions) has also been observed in the case of ruby lasers [31, 61].

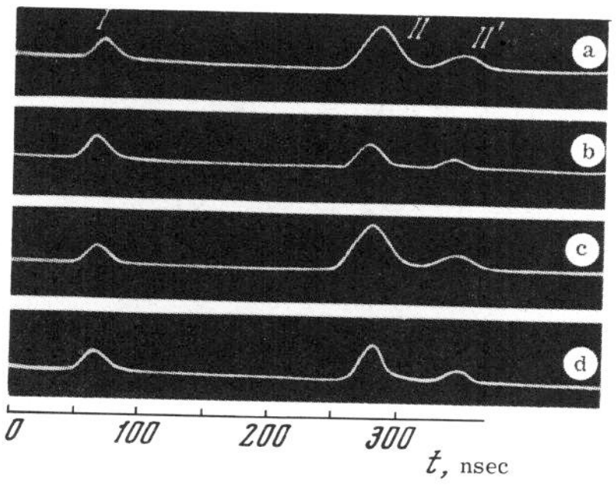

Fig. 10. Oscillograms of a giant pulse before (I) and after (II) a cell containing a bleachable solution (the appearance of pulse II' is caused by the reflection of signal II from the photocell FÉK-I due to mismatching of the active and wave impedances). a) Cell absent; b) neutral filter with transmissivity 0.5 in place of the cell; c) insertion of cell with $T_0 \sim 0.1$; d) insertion of cell with $T_0 \sim 0.001$.

Later investigations showed that from one to three axial modes could be excited with the particular experimental arrangement used here. In the first case mode interaction could not occur, so that the pulse had a regular form. With the excitation of two or three modes interaction of this kind could occur, and the pulse form could depart appreciably from regularity. A typical giant-pulse oscillogram used for the measurement of t_1 and t_2 is shown by way of example in Fig. 11. It clearly reveals the fact that the pulse has a skewed shape, i.e., the leading edge is steeper than the trailing edge. Since the values of the pulse width in [56] are given in units of the photon lifetime t_{ph}, we determined this quantity for the purpose of numerical comparison.

Under our experimental conditions the losses $\varkappa$ were determined by the losses due to the emission of radiation through the mirror R_2, i.e., $t_{ph} = 2L/[c(1-R_2)]$. For all four of the neodymium glass rods used we measured the total pulse width $t_1 + t_2$ and the ratio t_2/t_1 for various values of the transmissivity T_0. The results are summarized in Table 1, in which, due to the

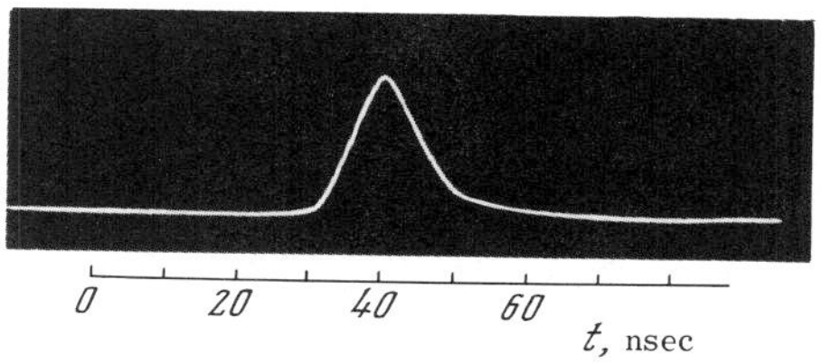

Fig. 11. Oscillogram of a giant pulse for a shutter transmissivity $T_0 = 0.4$ (sample dimensions 8 mm diameter × 130 mm).

TABLE 1

T_0	N_i/N_p	ln (N_i/N_p)	t_1+t_2, nsec	$\tau_1+\tau_2$	T_0	N_i/N_p	ln (N_i/N_p)	t_1+t_2, nsec	$\tau_1+\tau_2$
1.00	1,00	0.00	430	77	0,73	2.04	0.71	20	3.3
0.97	1,09	0.09	95	17	0.66	2.37	0.86	18,5	3,04
0.93	1.21	0.195	57	10.02	0.53	3.08	1,13	17.5	2,87
0,90	1.35	0.30	57	9.3	0,42	3,52	1.26	10.5	1,87
0.86	1.44	0.365	41	7,3	0.39	3.76	1,325	11	1,96
0.83	1.61	0,475	30	4.9	0,34	4.17	1.43	9.5	1.70

Notes: 1) The calculations were carried out for $R_1 = 0.99$, $R_2 = 0.51$ to 0.55 (the mirrors had to be changed due to the rapid hole burning in them at large powers), L = 41 to 42 cm, and $t_{ph} = 5.5$ to 6.1 nsec. 2) The results for $T_0 = 1.00$ are given for the case of the pure solvent.

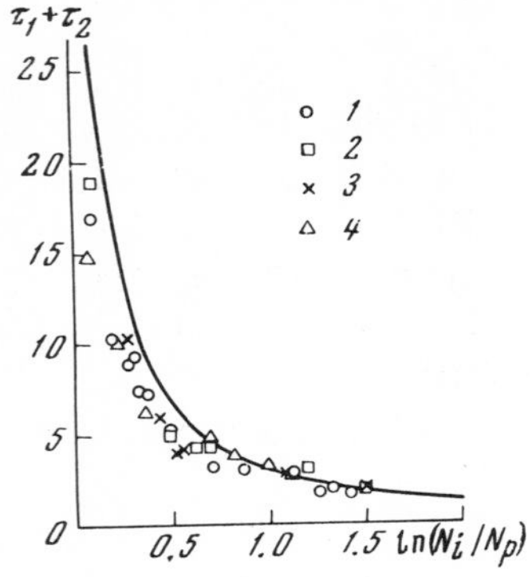

Fig. 12. Reduced giant pulse width $\tau_1 + \tau_2$ versus population inversion coefficient. Solid curve: according to theoretical calculations of [56]; 1) sample 8 mm diameter ×130 mm; 2) 9.3 mm diameter × 108 mm; 3) 12 mm diameter × 115 mm; 4) 15 mm diameter ×120 mm.

absence of any essential disparity between the results for different samples, only the results for one sample (8 mm diameter × 130 mm) are given as an example.

In order to facilitate the comparison of these results with the results of the theoretical calculations, in Fig. 12 we have plotted the theoretical dependence of the reduced pulse width $\tau_1 + \tau_2$ on ln n'_{ai} according to [56], along with the experimental points for all the samples investigated. As evident from Fig. 12, in the range of values of n'_{ai} from ~ 1.8 to 4.5 (interval of ln n'_{ai} from 0.6 to 1.5) the agreement between experiment and theory is reasonably good. But in the range of smaller values of n'_{ai} (0.8 to 1.1) the experimental points fall below the theoretical curve by values exceeding the possible experimental error.

Simultaneously with the deviation of the experimental values of the pulse width from the values predicted by theory [56], in the transition to the interval of $n'_{ai} < 1.4$ a qualitative change is noticed in the shape of the giant pulse in that the trailing edge becomes steeper relative to the leading edge (Fig. 13). This effect is qualitatively more noticeable from Table 2.

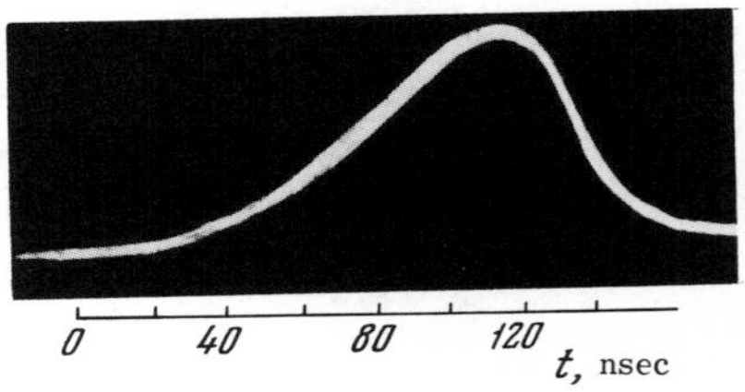

Fig. 13. Oscillogram of a giant pulse for T_0 = 0.92 (sample 12 mm diameter × 115 mm).

TABLE 2

T_0	n'_{ai}	t_2/t_1	
		Experiment	Theory [56]
0.98—0,97	1.07—1.09	0.97±0.02	1.02
0,94—0.92	1.21—1.31	0.77±0.07	1.03—1,08
0.91—0,89	1,31—1,33	0.86±0.03	1.08—1.10
0,87—0,85	1.42—1,53	1,10±0.05	1,11—1.13
0.83—0.79	1.61—1,78	1,11±0,05	1,14—1,18
0,74—0,72	1.99—2.04	1.20±0.07	1.20—1,24
0,69—0.66	2.09—2.37	1.35±0.10	1.24—1.30
0,56—0,50	2.70—3,28	1.45±0.10	1.38—1,48
0.42—0,34	3.52—4,45	1.44±0.20	1.50—1.70

This table gives the averaged experimental values of the ratio of the duration of the trailing to the leading edge of the giant pulse $(t_2/t_1)_{exp}$; the averaging is carried out over the data for different samples in a certain (indicated) interval of T_0. Also shown in the same table for comparison are the theoretical values of t_2/t_1 from [56].

The foregoing experiment thus exhibited good agreement between the values of the giant pulse width calculated theoretically for the model of an ideal shutter and measured directly from experiments with a neodymium glass laser using a passive shutter, for the interval of values of $n'_{ai} \gtrsim 1.8$, i.e., specifically for the most important interval from the point of view of generating short pulses with a high peak power.

The observed discrepancy between theory and experiment in the interval of small values of n'_{ai} is attributable to the fact that the shutter treated in [56] was an ideal version with instantaneous switching. However, this model can be applied to a bleachable filter only when certain conditions are met. They include, specifically, the requirement of fast [relative to the processes whose influence was neglected in the formulation of the system (5)] and complete bleaching of the filter, the latter condition being equivalent to the requirement of a large absorption cross section for the molecules of the filter relative to the induced-transition cross section for the active particles and a large lifetime for the upper level of the filter molecules relative to the giant pulse width. All of these conditions cannot be met for small values of n'_{ai}, in which case $t_1 + t_2$ is increased by an order of magnitude (relative to the same quantity for large n'_{ai}) and the photon density is decreased by three or four orders (we were able to estimate the order of variation of this quantity from the variation in the number of neutral attenuating filters placed in front of the photocell). In this case, therefore, it is required to analyze a more complete set of equations, taking into account the kinetics of the processes inherent in a bleachable filter (see, e.g., [62]).

With regard to the good agreement between experiment and the theory of an ideal shutter in the interval of $n'_{ai} \gtrsim 1.8$, later theoretical studies in which the processes involved in a laser with a passive shutter are analyzed have shed light on the quantitative aspect of this agreement. In these studies [32, 62-65] the processes in the laser are described by a set of three equations, i.e., by one more than in [56], the additional equation describing the difference between the populations of the levels of particles of the passive shutter:

$$\frac{d\Phi}{dt} = \left(\frac{\alpha l_a}{t_c} - \frac{\alpha_p l_p}{t_c} - \frac{1}{t_{ph}}\right)\Phi,$$
$$\frac{dN}{dt} = -\frac{2\alpha l_a}{t_c}\Phi, \qquad \frac{dM}{dt} = -\frac{2\alpha_p l_p \Phi}{t_c} - \frac{M_0 - M}{t_p}, \tag{9}$$

where $\alpha_p = \alpha_{p0}(M/M_0)$ is the absorption coefficient in the passive shutter (α_{p0} is the absorption coefficient in the absence of emission) and l_p is the working length of the cell forming the passive shutter. This elaboration of the set of equations makes it possible to account for the influence of the passive shutter parameters M, σ_p and t_p on the characteristics of the emitted giant pulses. The main results obtained in these studies by solution of the given system (9) (the solution was carried out by numerical methods on a computer) may be summarized as follows:

1. As $\tau_p = t_p/t_{ph} \to \infty$, for each value of $\sigma > n'_{ai}/(n'_{ai} - 1)$ there is an $(n'_{ai})_0$ above which, i.e., for $n'_{ai} > (n'_{ai})_0$, all characteristics of the giant pulse scarcely differ from the same characteristics in the ideal shutter case [32].
2. For $\sigma \sim 10^4$ the influence of the decay time of the upper level for particles of the shutter on the peak power characteristic of the giant pulse can be neglected for $n'_{ai} > 2$, down to $\tau_p \gtrsim 0.01$ [63].

3. In general, for the description of the characteristics of the pulse the following independent parameters are required: n'_{ai}, σ, and τ_p. For $\sigma > 200$ the characteristics of a laser with a passive shutter can be completely described by just two parameters: n'_{ai} and $\sigma\tau_p$ [65].

In light of these results we can estimate the limits of applicability of the theory of [56] to our situation. We adopt the following values of the parameters: $\sigma_a \approx 10^{-20}$ cm^2 [66]; $\sigma_p \approx 10^{-16}$ cm^2; $t_{ph} \approx 6 \cdot 10^{-9}$ sec; $t_p \approx 1.5 \cdot 10^{-10}$ sec [21], whereupon $\sigma \approx 10^4$ and $\tau_p \approx 2.5 \cdot 10^{-2}$. According to 2), the influence of the filter characteristics in this case can be neglected for $n'_{ai} > 2$; this result has been confirmed in our study.

CHAPTER II

SPECTRAL COMPOSITION OF SOLID-STATE LASER EMISSION AND MODE DISCRIMINATION BY VARIOUS CAVITY ELEMENTS

§1. Spectral Composition of Solid-State Laser Emission

The simplest solid-state laser embodies an active medium placed in a resonator cavity. In the earliest experiments on the generation of radiation in such lasers the cavity normally comprised two mirrors (metal or dielectric) deposited directly on the suitably finished (in the shape of spheres of definite radius or parallel planes) end faces of the active rod. It is understandable, then, that the first theoretical studies on the modal composition of the radiation in such a cavity [67-70] should presume the presence of only two mirrors. According to these studies the emission spectrum of the laser must consist of separate, equally spaced components, or modes (here and elsewhere we are concerned only with axial modes, i.e., TEM_{00p} modes in the classification of Fox and Li [67]), the distance between which (in cm^{-1}) is determined by the optical length of the cavity: $\delta\nu_m = 1/2L$.

However, the modal character of the laser emission spectrum only determines its fine structure. The general configuration of the solid-state laser spectrum must be determined by the frequency dependence of the gain. Consequently, in the simplest case the emission spectrum of the solid-state laser must represent a narrow luminescence line (the order of the narrowing depends on the degree of regeneration) with a fine structure of equidistant modes. The actual situation is far more complex, and the emission spectrum of real lasers often has a more complicated structure.

Present-day laser designs are enormously complicated, first of all, by the increase in the number of cavity elements (external mirrors, shutters, polarizers, etc.). All of these elements generally have plane-parallel surfaces and are placed inside the cavity (to reduce losses and facilitate tuning), usually parallel to the principal mirrors. The plane-parallel layers formed by the mirrors, end faces of the rod, and other surfaces, however, act as Fabry–Perot interferometers with a wavelength-dependent transmissivity. Theoretical calculations of the spectrum of possible modes of this type of cavity [71-74] show that it can be considerably more intricate than in the case of the two-mirror cavity.

Merely knowing the spectrum of possible modes of the laser cavity and the Q-values for each one does not suffice to answer the question of what the emission spectrum of the laser will be like in a particular emitting state. The reason for this is that the formation of the emission spectrum can differ to a very great extent for different operating regimes of the laser, i.e., in single- or multiple-peak free oscillation, giant-pulse emission with modulation by a Kerr cell or a passive shutter, etc. And besides the fact that each of these regimes is characterized by its own characteristic oscillation buildup time, the formation of the spectrum in some of these regimes might evolve other than by independent growth of the intensity of the individual modes, namely with interaction induced by the nonlinear interaction of both the active medium itself and any other element of the cavity system. The latter can also produce qualitative changes in the resultant spectrum of the laser. In addition to all of the foregoing, there are effects which are very difficult to take into strict account. Typical of such effects are various types of inhomogeneities (of the active medium, pumping, cavity elements), thermal effects, etc. Finally, it is essential to realize that already in the first investigations of the emission spectra of lasers incorporating passive shutters [26, 37, 75] a severe narrowing was observed in the giant-pulse spectrum by comparison with the free-oscillation spectrum. This change in the spectrum was directly attributed by some authors [37] to the action of the passive shutter, to such a convincing extent that the term "selective action of the passive shutter" was even coined.

The existence of the factors listed above indicates that the thorough investigation of the emission spectrum of lasers with a passive shutter must include the following aspects: first, an analytical determination, if possible, of the spectrum of cavity modes and their Q-values; second, an experimental investigation of the resultant emission spectrum; third, a study of the oscillation-emission kinetics in order to ascertain the causes of possible inconsistencies between that spectrum and the analytical version; and, fourth, a clarification of the role of various elements in the formation of the spectrum.

§2. Mode Discrimination in Neodymium Glass Lasers with a Passive Shutter Due to Fresnel Reflection at the Ends of the Active Rod

The influence of Fresnel reflection at the end faces of the active rod on the spectrum of excited modes as manifested in a variation of the beat frequency between these modes was first noted in our paper [18]. This influence was attributed to the discrimination of certain axial modes of the main cavity (i.e., the cavity formed by external mirrors) and depended on the position and tuning of the active rod. The laser in this case operated in the passive Q-switching regime. A neodymium glass laser with external plane mirrors R_1 and R_2 with reflectivities of 0.98 and 0.65, respectively, was used in the experiment. A cylindrical KGSS-7 glass rod with plane-parallel ends, 12 mm in diameter and 120 mm long, was used. The Q-switching operation was realized by means of a bleachable liquid filter with a solution of No. 7 dye in nitrobenzene. The transmissivity of the filter at the emission wavelength $\lambda = 1.06\,\mu$ was normally $T_0 = 0.7$ to 0.75. The cell containing the solution was set up near R_1, and the inclination of its plane relative to the cavity axis was $\sim 1°$. The emission was detected with an FÉK-09 photocell connected to the CRT of an S1-11 oscilloscope. The optical pathlength L of the cavity could be varied from 40 to 320 cm.

In the first part of the experiment the active rod was set up near R_1 (the distance from the center of the rod to R_1 was $L_1 \approx 20$ cm), and the cavity length was varied by displacement of R_2. As L was increased from 40 to 150 cm the width of the giant pulse, $t_1 + t_2$, increased from ~ 25 nsec (L = 40 cm) to between 70 and 90 nsec (L = 150 cm), the shape of the pulse recorded by our instrumentation remaining fairly smooth. Beginning with $L \gtrsim 150$ cm, however, the time scan of the pulse acquired a regular (periodic) structure with period T_{str} equal, within the experimental error limits ($\sim 10\%$), to the reciprocal of the frequency separation of the adjacent modes: $T_{str} = 2L/c$ (Fig. 14). The percentage modulation varied from one pulse to the next, but on the average it increased with L, attaining close to 100% for large lengths ($L \approx 300$ cm).

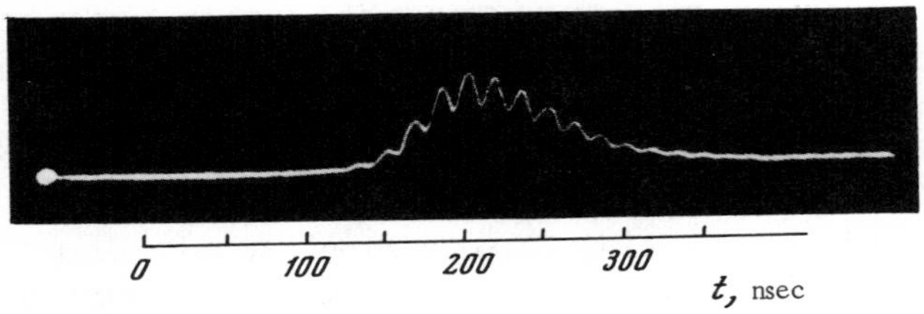

Fig. 14. Oscillogram of a giant pulse emitted by a neodymium glass laser with a passive shutter ($T_0 \approx 0.7$; $L = 228$ cm).

In the second part of the experiment the cavity length was held constant at L = 320 cm, and the distance L_1 was varied between 20 and 300 cm. The ends of the rod in this case remained parallel to the cavity mirrors (with less than 10" error). It turned out that the modulation frequency also varied with displacement of the rod, the variation in this case acquiring a jumplike behavior: in a certain interval of values of L_1 the modulation frequency remained constant, then with a further variation of L_1 it jumped suddenly. The intervals of variation of L_1 in which the maximum percentage modulation was attained at a corresponding constant modulation frequency f_{exp} are given in Table 3. Also shown in the table are the closest-spaced beat frequencies between cavity modes: $f_m = m(c/2L)$, where m = 1, 2, 3, ... Oscillograms of pulses for various values of the distance between the center of the active rod and R_1 are shown in Fig. 15.

As evident from Table 3, with the neodymium glass rod situated at distances $L_1, L_2 > 60$ cm from R_1 or R_2 the observed modulation frequencies coincide (within the experimental error limits) with one of the values of f_m (m = 2, 3, 4). For $L_1, L_2 < 50$ cm the modulation frequency is equal to f_1. It is important to note that in the L_1-intervals indicated in Table 3 the maximum percentage modulation at the frequency f_m is observed for values of L_1 approximately equal to $L_{1m} = L/m$. For values of L_1 differing by 15 to 20 cm from L_{1m} the stability of the occurrence of a particular multiple frequency f_m decreases, i.e., first f_2, then f_3 ($L/2 < L_1 < L/3$) or f_3 and f_4 ($L/3 < L_1 < L/4$) appear. For $L_1, L_2 < 50$ cm higher modulation frequencies f_m (m = 5, 6, . . .) are not observed, but the fundamental frequency f_1 is observed. The character of the modulation in this case is rather extraordinary, i.e., the emission has the form of equidistant narrow spikes, whose widths are considerably smaller than the distance between them (Fig. 15a). Also, the percentage modulation decreases as the frequency increases (Figs. 15a-15d).

Our observed variation of the laser emission modulation frequency with the position of the active rod was not only detected in the Q-switching regime, but also in the free-oscillation region. Under the same experimental conditions, but without the bleachable filter, we investigated the time behavior of the emission of an individual free-oscillation peak. It also turned out to have a periodically modulated intensity, the modulation frequency exhibiting the same

TABLE 3

L_1, cm	f_{exp}, MHz	f_m, MHz	m	L_1, cm	f_{exp}, MHz	f_m, MHz	m
20—50	44.5	46.9	1	205—230	134	140.7	3
60—85	180	187.6	4	235—260	180	187.6	4
95—115	133	140.7	3	270—300	44	46.9	1
150—170	90	93.8	2				

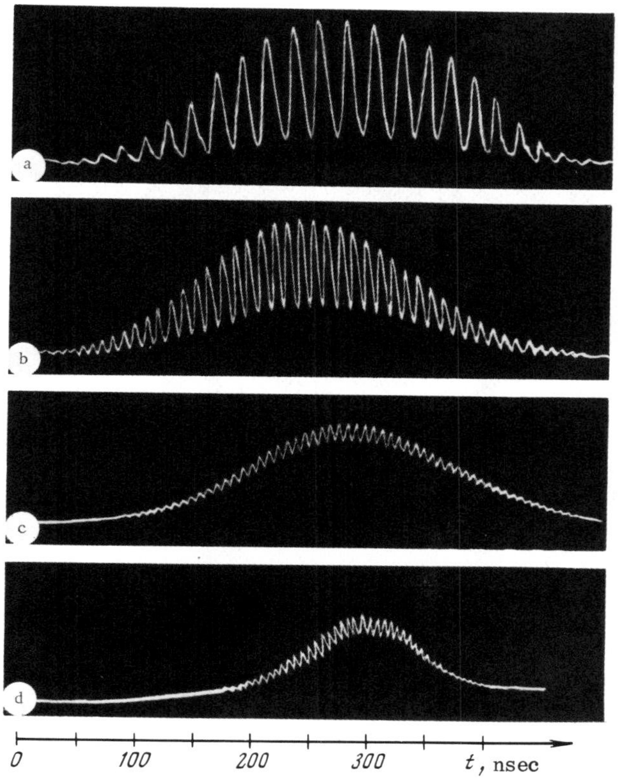

Fig. 15. Oscillograms of giant pulses emitted by a neodymium glass laser with a passive shutter (L = 320 cm). a) L_1 = 30 cm; b) L_1 = 161 cm; c) L_1 = 107 cm; d) L_1 = 74 cm. a-c) T_0 = 78%; d) T_0 = 72%.

behavior as in the giant-pulse case (Fig. 16). The percentage modulation in this case, however, is considerably lower.

The modulation of the output signal at frequency multiples of c/2L is a consequence of the modal composition of the emission spectrum, a fact that had already been utilized earlier for the investigation of the modal composition of ruby lasers [76, 77]. We were the first to observe beats of this type for neodymium glass lasers in the Q-switching regime. Two aspects were noticed in the observed effect. First, the jumplike variation of the modulation frequency with variation of the position of the active rod is the same in both regimes, and, second, the percentage modulations differ sharply in the two regimes. The first consideration is explained by the fact that the variation of the modulation frequency is based on the same mechanism of variation of the modal composition of the emission spectrum. The second consideration was shown by later investigations to be related to the self-mode-locking effect in lasers using a passive shutter. We now give more careful attention to the fact that, prior to our work, beats had not been observed in the output signal of a neodymium glass laser in the free-oscillation regime.

The emission of lasers may be regarded in most cases with reasonable approximation (disregarding nonaxial modes TEM_{mnp} with m, n ≠ 0) as consisting of a set of plane waves

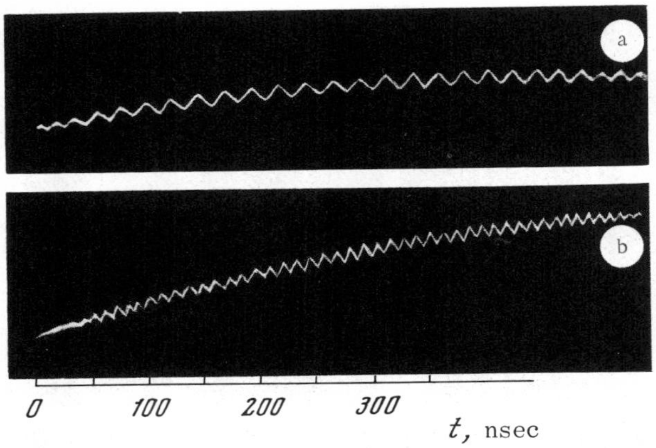

Fig. 16. Oscillograms of a free-oscillation peak from a neodymium glass laser (L = 320 cm). a) L_1 = 20 cm; b) L_1 = 160 cm.

(modes) whose frequencies are equidistant and whose phases are random variables.* The latter fact implies the absence of locking. In the registration of this type of radiation with a square-law detector the constant component of the signal will have superimposed on it beats with frequencies that are multiples of the mode separation c/2L. For a random distribution of phases of the individual modes the form of the modulation and its percentage criterion will also vary randomly from one burst to the next. For a very large set of modes, however, the random character of the phase distribution will tend to smooth out the beats, owing to the high probability in this case that for any pair of modes with a beat frequency f_m there will be another pair of modes with the same beat frequency but opposite in phase.† But if there are only a few excited modes this smoothing effect cannot occur. From the point of view of the foregoing considerations, therefore, the observation of beats between axial modes requires (in addition to the separation of these modes by the suppression of TEM_{mnp} modes with large m and n) that the number of excited modes be decreased. A certain confirmation of this point of view may be found in the fact that prior to publication of our paper [18] the modulation of solid-state laser emission at the beat frequency between modes had only been observed in ruby lasers in the free-oscillation regime [76, 77, 79, 80] and in Q switching by means of a rotating prism [81]; the obvious reason for this situation was the narrower width of the ruby luminescence line (by comparison with a neodymium glass laser).

In our experiment conducted to observe the radiation modulation of the free-oscillation peak in a neodymium glass laser the stated requirements were met by a proper choice of cavity arrangement. The latter (L ≈ 300 cm) was such as to prevent the excitation of angular modes having large angular indices, because the losses for these modes are particularly high in the given arrangement. Moreover, the use of mirrors on plane-parallel substrates and an active rod with end faces perpendicular to the cavity axis produced a sizable reduction in the number of excited modes by virtue of discrimination.

*See [78] with regard to the random phase distribution of the individual modes of solid-state lasers in the free-oscillation regime.

†We are concerned here with unsynchronized beats, i.e., beats between modes with random phases. In the synchronization case (mode locking, which is realized, for example, in a laser with a passive shutter) increasing the number of modes not only does not detract from the occurrence of beats, it can even have the converse effect of reinforcing them.

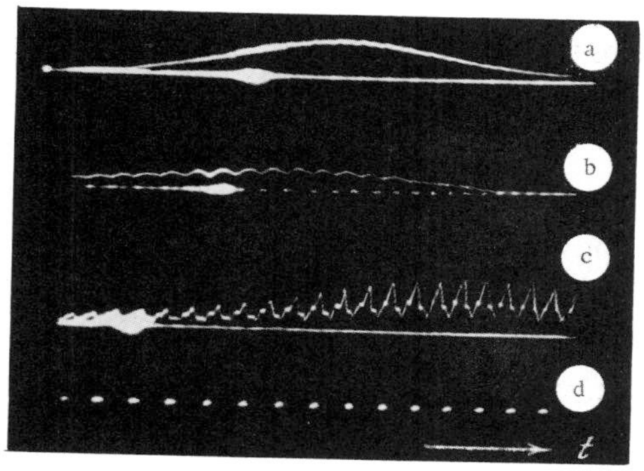

Fig. 17. Oscillograms of free-oscillation peaks emitted by a neodymium glass laser (L = 107 cm). a) $\sim 2 \cdot 10^3$ modes excited; b) 300 to 600 modes excited; c) about 10 modes excited; d) 10-nsec time marker.

We set up a special experiment designed to exhibit the influence of the number of excited modes on the percentage modulation. The neodymium glass laser operated in the single-peak free-oscillation state. The need for this operating regime was dictated by the fact that concurrently with the observation of the time dependence of the radiation intensity we conducted spectral measurements, in which there was no provision for time resolution, but rather the integral pattern was investigated. The optical length of the laser cavity in this case was L = 107 cm. The active rod (15 mm diameter × 220 mm), its ends cut at the Brewster angle, was placed in the center of the cavity. Two diaphragms 3 mm in diameter were mounted near the ends of the rod for the suppression of nonaxial (angular) modes. The time-recording circuit consisted of an FÉK-09 photocell and I2-7 oscilloscope. By the proper choice of such cavity elements as the external mirrors and internal mode selectors we were able to vary the total number of excited modes in the laser emission spectrum (with complete control) from $\sim 2 \cdot 10^3$ to ~ 10. The time dependence of the free-oscillation peak in these cases had the forms shown in Fig. 17. As the oscillograms clearly reveal, the percentage modulation of the output signal depends very strongly on the number of modes excited. Thus, in the case of ~ 10 modes (Fig. 17c) the percentage modulation is greater than 50%, whereas with the excitation of $\sim 2 \cdot 10^3$ modes (Fig. 17a) there is virtually no modulation.

The occurrence of modulation frequencies which are multiples of $f_1 = c/2L$ with variation of the position of the active rod can be explained by the variation of the spectrum of excited axial modes due to the change of their Q-values. For definite rod positions the excited modes are not adjacent, but have a frequency separation mf_1, so that the modulation frequency is increased. Consequently, our observed variation of the modulation frequency is interpreted as the discrimination of certain axial modes with displacement of the active rod.

There is, of course, one other possibility for integer multiplication of the modulation frequency, even when all possible modes of the cavity are excited. This possibility has to do with the self-mode-locking effect in some nonlinear cavity element. In the given case the element in question would have to be the active rod, because it is the position of the latter that affects the observed variation of the modulation frequency. However, since the frequency variation occurred only in the event of strict parallelism between the rod end faces and the cavity mirrors, whereas with the end faces tilted relative to the plane of the mirrors by an angle > 2' the modulation frequency was equal to f_1 and independent of the rod position, this possibility is precluded.

The discrimination of certain axial modes can be explained by considering the given system as a complex cavity formed by the external mirrors R_1 and R_2 and the ends of the neodymium rod. In our paper [82] we accounted for the observed regularities on the basis of a theoretical analysis of the indicated type of cavity formed by four reflecting surfaces, and we gave some experimental results by which it is possible to assess the role of various factors in the observed effect. The problem of determining the eigenfunctions (in the form of "plane" waves) of the cavity with four reflecting surfaces in general form has already been solved a number of times (see [72-74]). Its solution based on the assumption of mirrors of infinite extent, i.e., disregarding diffraction losses, entails a determination of the eigenvalues of the wave number k. The number k is complex: $k = k' + ik''$, where $k' = 2\pi/\lambda$ and k'' is the loss factor. In our problem the equation for k has the form

$$\exp(-2ikL) = \frac{1 - rr_1 \exp(-2ikl_1)}{r \exp(2ikl_1) - r_1} \frac{1 - rr_2 \exp(-2ikl_2)}{r \exp(2ikl_2) - r_2}, \tag{10}$$

where r_1, r_2, and r are the reflectivities for the amplitudes of electromagnetic waves at the cavity mirrors and ends of the rod, respectively, and l_1 and l_2 are the distances of the ends of the rod from the corresponding cavity mirrors. If we assume that $1 - R_1 \ll 1$, $1 - r_2 \ll 1$, and $r \ll 1$, we obtain a solution of Eq. (10) for k in the form of a sum $k = k_0 + k_1$, where k_0 is the solution for the case $r = 0$, i.e., in the zeroth approximation, and k_1 represents the first approximation. The expressions for $k_0 = k_0' + ik_0''$ and $k_1 = k_1' + ik_1''$ have the following form in this case:

$$k_0'L = s\pi \tag{11}$$

(s is an integer);

$$k_0''L = \frac{1}{2}[(1 - r_1) + (1 - r_2)]; \tag{12}$$

$$k_1'L = -r(\sin 2k_0'l_1 + \sin 2k_0'l_2); \tag{13}$$

$$k_1''L = r\left\{\left[(1 - r_1) - (1 - r_1 + 1 - r_2)\frac{l_1}{L}\right]\cos 2k_0'l_1 + \left[(1 - r_2) - (1 - r_1 + 1 - r_2)\frac{l_2}{L}\right]\cos 2k_0'l_2\right\}. \tag{14}$$

Equations (11) and (12) are the usual conditions governing the frequency and losses of axial modes in the case of the "empty" cavity. Equations (13) and (14) determine the frequency shift and correction to the axial mode losses due to insertion of the two extra reflectors (ends of the rod) with amplitude reflectivity r into the "empty" cavity. The significant fact here is that, according to (13) and (14), the magnitude and sign of both corrections depend on both r and the ratio between l_1 and l_2 and are different for different axial modes (i.e., for different integers s). This means, in particular, that with the insertion of the two additional mirrors into the "empty" cavity the equal spacing of the modes is upset. A calculation of the relative frequency shift k_1'/k_0' of the mode for our specific case has shown, however, that this effect is considerably less than the possible experimental error (~ 10%) and therefore went undetected.

Let us examine in closer detail the correction for the losses and its dependence on the ratio between l_1 and l_2 and on the mode number s. Inasmuch as the quantity $k'' = k_0'' + k_1''$ is the loss factor, a minus sign on the correction for k_1'' connotes a reduction in losses, i.e., an increase in the Q of the complex cavity for the corresponding mode, whereas, conversely, a plus sign implies a reduction in the Q. For comparison with the experimental results we calculated the correction $k_1''L$ as a function of the mode number s for the cases $L_1 = L/m$ (m = 2, 3, 4, 5), i.e., for positions of the active rod within the cavity such that the maximum modulation amplitude is observed at the corresponding frequency. The case $L_2 = L/m$ differs from the one in question only by the numerical factor associated with $k_1''L$, the nature of the dependence on s remaining unchanged. The results of the calculations are presented graphically in Fig. 18.

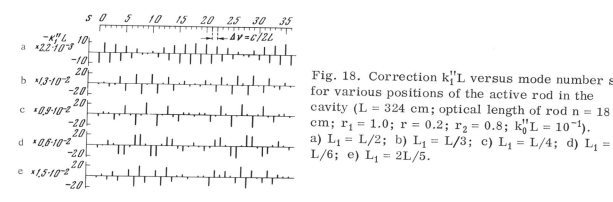

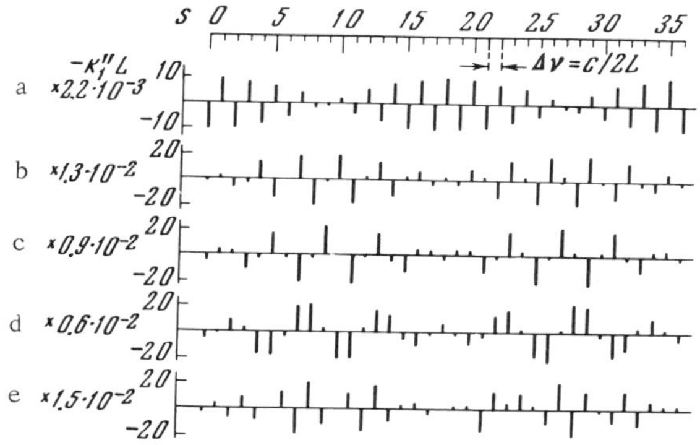

Fig. 18. Correction $k_1''L$ versus mode number s for various positions of the active rod in the cavity (L = 324 cm; optical length of rod n = 18 cm; $r_1 = 1.0$; $r = 0.2$; $r_2 = 0.8$; $k_0''L = 10^{-1}$). a) $L_1 = L/2$; b) $L_1 = L/3$; c) $L_1 = L/4$; d) $L_1 = L/6$; e) $L_1 = 2L/5$.

According to Fig. 18a, in the case $L_1 = L/2$ the minimum losses occur for modes with s = 1, 3, 14, 16, 20, etc. In this case the minimum frequency interval between excited modes, which determines the modulation frequency of the output signal, is equal to 2(c/2L). In our experiment we did in fact observe the frequency $f_{exp} = f_2 = 2(c/2L)$ in this case. In the case $L_1 = L/3$ (Fig. 18b) the predominantly excited modes were s = 4, 7, 10, 13, etc. The beat frequency must change accordingly: $f_{exp} = f_3 = 3(c/2L)$. The same argument applies to the case $L_1 = L/4$ (Fig. 18c). Consequently, the integer multiplication of the modulation frequency with displacement of the active rod inside the cavity is fully accounted for on the basis of an analysis of the Q-values of the modes of a four-mirror cavity.

We also calculated the correction $k_1''L$ as a function of s for intermediate positions of the active rod in the cavity, i.e., for values of $L_1 \neq L/m$: $L_1 = 2L/5$, $L_1 = 2L/7$, etc. The case $L_1 = 2L/5$ is shown for illustration in Fig. 18e. In this case the frequency intervals between modes with a large negative correction are not constant. Thus, the interval between modes with s = 3 and 6 and between those with s = 8 and 11 is equal to 3(c/2L), while between modes with s = 6 and 8, 11 and 13 it is equal to 2(c/2L). The modulation frequency in this case will depend very strongly on the relative position of the maximum of the spontaneous emission line and spectrum of eigenfrequencies of the cavity, and this relative position can change, depending on the external conditions (for example, heating of the rod). This particular instability of the modulation frequency was indeed observed experimentally.

It is important to note that with a reduction in L_1 the difference between the values of the correction $k_1''L$ for the mode with a maximum negative value and for one of the adjacent modes diminishes rather quickly and can in principle lead to a situation in which for sufficiently small L_1 not only the modes having a maximum negative value of the correction $k_1''L$ (i.e., modes s = 8, 13, 23, 28, etc., in Fig. 18d), but also adjacent modes (s = 7, 14, 22, 29, etc.) will be strongly excited. The modulation frequency of the output signal in this case will be determined by the frequency interval between adjacent modes, i.e., c/2L. We did observe such a dependence experimentally, i.e., beginning approximately with a value of $L_1 < L/6$, the modulation frequency was independent of the position of the rod and was equal to $f_{exp} = f_1$.

As apparent from Fig. 18, the absolute value of the correction for the Q-values of the different modes is of the order 10^{-2} or 10^{-3}, i.e., is fairly small. But the fact that this correction suffices for the dominant excitation or suppression of certain modes attests to a certain specific quality inherent in the buildup of the giant-pulse emission spectrum in a laser with a passive shutter. This specific aspect, which is also manifested in a radical narrowing of the giant-pulse spectrum in passive switching by comparison with the free-oscillation spectrum was first brought to attention in [26, 75].*

According to [83], in which this problem is subjected to theoretical analysis, the photon density buildup required in the cavity for initiation of the filter bleaching process occurs over a rather large time interval, about $7 \cdot 10^2 \cdot t_c$, where t_c is the round-trip photon transit time across the cavity (the estimate is given for phthalocyanine). As a result, even a minute initial discrepancy between the Q for two particular modes can lead to a very pronounced difference between the intensities of these modes at the instant of bleaching (and, hence, at the instant of giant-pulse emission).

A quantitative order-of-magnitude estimate of the required initial difference in the Q-values of two modes whose intensity will differ by one order of magnitude in the giant pulse with passive Q switching is $\sim 3 \cdot 10^{-3}$ [83].

In the case of an ideal shutter with instantaneous switching, on the other hand, a similar calculation gives a value of about $30t_c$ for the giant-pulse buildup time. Accordingly, the intensity ratio of any modes in the giant pulse that differ by one order of magnitude with passive switching will differ only slightly from unity in this case. Thus, when an instantaneous switch is used, the initial Q discrimination of certain modes will be much less pronounced.

We decided in light of these considerations to compare the spectral composition of laser emission with passive switching and some other kind of switching that could be regarded as nearly ideal. For the latter, on the basis of design simplicity, we chose a shutter based on a rotating prism with total internal reflection [2, 84]. The buildup time of the giant pulse with this type of shutter can be assumed approximately equal to the time required for the prism to rotate through an angle equal to the angle of divergence ψ of a single giant pulse emitted by the laser. Inasmuch as this angle was equal to $(1 \text{ to } 2) \cdot 10^{-3}$ in our case and the rotation speed of the prism was 30,000 rpm, the pulse buildup time in this system was of the order $(3 \text{ to } 7) \cdot 10^{-7}$ sec = $(50 \text{ to } 120)t_c$. It is permissible, therefore, to regard this type of shutter as reasonably close to ideal.

The emission spectra were investigated with a diffraction spectrograph, whose optical scheme is illustrated in Fig. 19. As the figure indicates, the spectrograph consisted of a UF-85 autocollimation tube with focal length f = 1300 mm and an echelette grating (200 lines/mm) operating in the eighth order (the angle of incidence of radiation on the grating is $\varphi_d \approx 58°$,

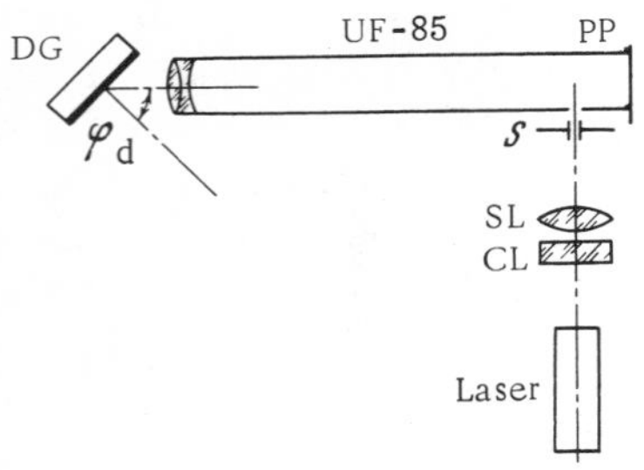

Fig. 19. Optical schematic for investigation of the emission spectrum of a neodymium glass laser. DG) Echelette diffraction grating (200 lines/mm); S) entrance alit; PP) plane of photographic plate; SL) spherical lens; CL) cylindrical lens.

---

* It is important to note that this kind of comparison of the emission spectra in the two regimes is often made incorrectly, in that the spectrum of a single giant pulse is compared with the integral spectrum of multiple-spike free oscillation, but the complex nature of the latter spectrum is specifically attributable to the large number of peaks involved.

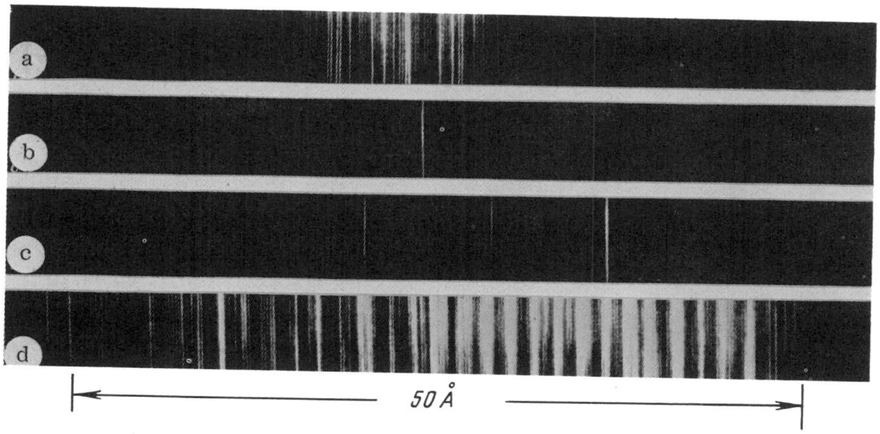

Fig. 20. Spectrograms of emission from a neodymium glass laser (sample 12 mm diameter × 120 mm). a) Emission of a single giant pulse with a rotating-prism switch; b) a single giant pulse with passive switching (T_0 = 45%, pumping at a maximum of 10% above threshold); c) three giant pulses with passive switching (T_0 = 45%, pumping at 70 to 80% above threshold); d) free oscillation at seven times the threshold level.

corresponding to the blaze angle from the small face of the rule line [85]). The linear dispersion of the instrument was $d\lambda/d \approx 2.5$ Å/mm, so that with an entrance slit width of 0.04 mm the spectrograph had a resolution of ~ 0.1 Å. A spherocylindrical condenser was used to ensure uniform coverage of the height of the entrance slit with the laser emission. The spectrum was recorded on infrared film.

Spectrograms of the emission from a neodymium glass laser in four regimes are presented in Fig. 20. In every regime a KGSS-7 glass rod (12 mm diameter × 120 mm) with an etched lateral surface, its end faces aligned parallel to the cavity mirrors with a maximum error of 10", was used as the active rod. The exit mirror was deposited on a plane-parallel substrate and had a reflectivity $R_2 \sim 0.6$. The emission spectrum in Fig. 20a corresponds to the case of a single giant pulse with switching by a rotating prism. The cavity length in this case was L = 90 cm. As the spectrograms show, the total width of the spectrum is about 25 Å, and the spectrum itself is made up of a large number of sharp lines of varying intensities, the widths of which are determined by the instrument function of the equipment. The spacing of the adjacent lines is equidistant and equal to $\delta\lambda_{str} \approx 0.20$ Å ($\delta\nu_{str} \approx 0.18$ cm^{-1}). This fine structure, as we explained, was a result of interference at the rotating prism, the action of which is equivalent to a plane-parallel plate, so that the period of the structure is intimately related to the height h of the prism (distance from its vertex to the entrance face): $\delta\nu_{str} = 1/4nh$, where n is the refractive index of the prism material.

The emission of a single giant pulse with passive switching (solution of No. 7 dye in nitrobenzene) is shown in Fig. 20b. The cavity had the same structure as before, apart from the replacement of the rotating prism with a mirror R_1 having a reflectivity $\gtrsim 0.99$ and a reduction in length to L = 70 cm. The cell containing the bleachable solution ($T_0 \approx 45\%$) was placed near R_1. The selective function of the rotating prism in this case was executed by the external surfaces of the cell windows (reflection at the boundary between the inside surface of the windows and the nitrobenzene can be neglected in this case). The emission of a single giant pulse was effected by control of the pumping. The emission spectrum in this case comprises a single line of width $\Delta\lambda \leqslant 0.1$ Å, which is determined by the instrument function.

The emission spectrum of the laser with passive switching in a regime in which several giant pulses are emitted in series (during a single pumping pulse) consists (Fig. 20c) of a corresponding number of individual narrow lines occupying an interval of about 25 Å.

The free-oscillation spectrum (solution removed from the cell) is shown in Fig. 20d for comparison. Note, however, that since the laser emission in this case represents a sequence of many spikes on the time scan, it is impossible to identify the spectrum with each of the individual spikes. The total width of the emission spectrum is approximately equal to $\Delta\lambda \approx 60$ Å, while the spectrum itself, as in the case of Fig. 20a, consists of a great many groups of sharp lines of varying intensities but equally spaced with a period $\delta\nu_{str} \approx 0.24$ cm^{-1}. This structure* is caused by interference at the exit mirror, and its period is therefore determined by the substrate thickness d_{sub} of that mirror: $\delta\nu_{str} = 1/2nd_{sub}$.

Inasmuch as the observed line width in the emission spectrum of a single giant pulse with passive switching was determined by the instrument function of the spectrograph, we decided to use a Fabry–Perot interferometer for a more detailed study of the structure of this line. We anticipated that the use of an interferometer with appropriate resolution would enable us to investigate the spectrum of excited axial modes and, hence, the discrimination effect inherent in the direct spectroscopic method. For reliable resolution of the adjacent axial modes by means of the Fabry–Perot interferometer at our disposal, which had a plate separation $d_{FP} = 15$ cm, we made the length of the cavity in this experiment equal to L = 60 cm. Since high-order transverse modes are generally excited in such a short cavity, we set up a diaphragm 3 mm in diameter near one end of the active rod in order to eliminate those modes. The KGSS-7 glass active rod (12 mm diameter × 130 mm) was positioned inside the cavity so that ($l_1 = 15$ cm) the correction $k_1'L$ would differ appreciably for adjacent modes, i.e., so that strong discrimination of these modes would occur with the rod in perfect alignment. In the experiment we simultaneously investigated the spectral composition of the emission of a single giant pulse (with the Fabry–Perot interferometer) and the time dependence of the output radiation (with an FÉK-09 photocell and CRT of an S1-11 oscilloscope) for two alignments of the rod relative to the plane of the cavity mirrors. In the first case the plane-parallel ends of the rod were tilted ~ 1° with respect to the plane of the mirrors, and in the second case they were parallel (to within 10").

The results of the experiment are shown in Fig. 21. As the interference patterns of Fig. 21a show, the giant-pulse spectrum in the case of the tilted rod consists of four components separated by a distance $\delta\nu_{str} = 0.83 \cdot 10^{-2}$ cm^{-1}, which concurs, within the experimental error limits (~ 5%), with the separation of the adjacent axial modes of the cavity, so that the total width of the spectrum is $2\Delta\nu_{osc} \approx 2.5 \cdot 10^{-2}$ cm^{-1}. The time scan of the giant pulse in this case (Fig. 21c) clearly reveals the intensity modulation at a frequency $f_1 = c/2L$ due to beats between adjacent modes. This small number of excited modes in the spectrum is a result of the discriminating action of the entire set of such cavity elements as the outer surfaces of the plane-parallel substrates, the plane-parallel walls of the cell, and end faces of the rod. The reflectivities and transmissivities of all these elements, due to interference effects in the plane-parallel layers, are periodic functions of the wavelength, the period being determined by the optical thickness of the corresponding plane-parallel layer. The Q equation contains all of these functions in product form, because the wavelength dependence turns out to be rather complex. However, as the number of plane-parallel elements is increased, the number of

* The reason that this structure is not observed in the giant-pulse spectrum in the case of switching by a rotating prism is clearly the fact that in the latter case the discrimination with the prism is stronger.

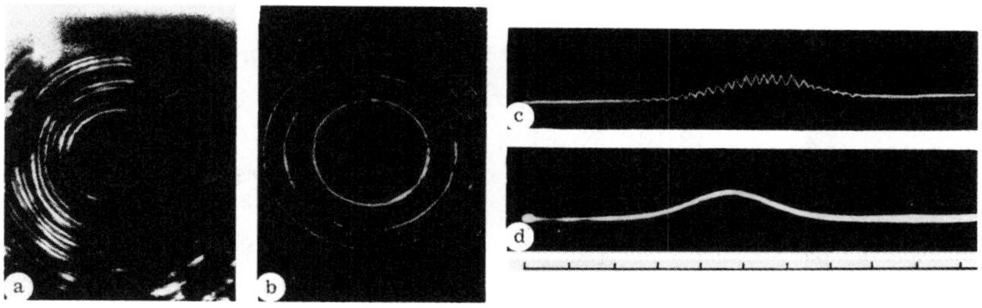

Fig. 21. Interference patterns of the emission spectrum (a, b) and oscillograms (c, d) of a giant pulse from a neodymium glass laser with a passive shutter ($L \doteq 60$ cm; $l_1 = 15$ cm; $nl = 20$ cm; sweep scale: 20 nsec/div). a, c) ends of active rod tilted $\sim 1°$; b, d) ends parallel to cavity mirrors.

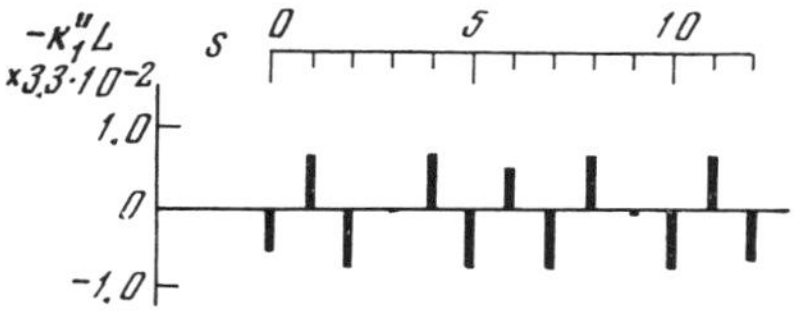

Fig. 22. Correction $k_1''L$ versus mode number s for a cavity with $L = 60$ cm ($nl = 20$ cm; $l_1 = 15$ cm; $r_1 = 1.0$; $r_2 = 0.8$; $r = 0.2$; $k_1'L = 10^{-1}$).

modes having the maximum Q within the span of the luminescence line decreases, and so also does the number of modes excited in the laser with a passive shutter.

In the case of the perfectly aligned rod the giant pulse has a smooth form (Fig. 21d), and its spectrum contains only one axial mode (Fig. 21b), whose width is determined in our case by the instrument width and does not exceed $2\Delta\nu_{osc} \leqslant 3.3 \cdot 10^{-3}$ cm^{-1}. Its width is actually limited, obviously, only by the pulse width (Fig. 21d) and in this case is $2\Delta\nu_{osc} \approx 1/[c(t_1 + t_2)] \approx 10^{-3}$ cm^{-1}. The excitation of only one axial mode in this case is easily explained by the results of a calculation of the correction $k_1''L$ for different modes according to Eq. (14), as shown graphically in Fig. 22. According to Fig. 22, modes with s = 1, 4, 8, 11, etc., with a minimum separation equal to $3(c/2L)$, should be the most strongly excited in this case. But since the width of the emission spectrum, even with the rod tilted, is no more than $2.5 \cdot 10^{-2}$ cm^{-1}, i.e., no more than four adjacent modes can be excited at one time, the intense excitation of only one mode is the most probable occurrence in the case of the perfectly aligned rod.

It is essential to note that besides the conceptual significance of the single-mode emission regime in a neodymium glass laser with a passive shutter (narrowing of the emission line by more than five orders of magnitude relative to the width of the luminescence line), it has tremendous practical value in broadening considerably the area of application of this type of laser.

§3. Selective Properties of Various Cavity Elements and Their Influence on the Emission Spectrum of Solid-State Lasers

As stated in §§1 and 2 above, the emission spectrum of a solid-state laser with external mirrors and other parallel reflecting surfaces inside the cavity can be fairly complex. Changing the relative positions of the elements with plane-parallel surfaces in such a cavity can have a significant influence on the spectrum of excited modes of the laser within the width of the luminescence line of the material of the active rod. This fact is clearly implied by Eq. (14), according to which a change in the distance from an external mirror to one of the reflecting layers (l_1 or l_2) by an order of $\lambda/2$ can alter not only the correction to the Q-values for certain modes, but also the sign of this correction, i.e., modes having the largest Q will have the smallest Q. Particularly large variations in the emission spectrum should be expected in the case of neodymium glass lasers (an example of such variations of the spectrum in the free-oscillation regime may be found in [86]), whose luminescence line width, according to [66, 87], is about 250 Å.

In our first investigations [18, 82] of the discrimination of excited modes due to Fresnel reflection at the ends of the active rod the total width of the emission spectrum of the neodymium glass laser with a passive shutter was considerably smaller (by about four orders of magnitude) than the luminescence line width. The same situation was observed in the case of a ruby laser with a passive shutter [26]. The results of these studies justified the hypothesis that the indicated difference could be attributed to the aggregate selective action of all the cavity elements. It was expected that if discrimination could be completely eliminated in the cavity the width of the emission spectrum would be increased and be determined solely by the width of the luminescence line of the active medium and pulse buildup processes.

In order to shed light on this problem we undertook some systematic investigations of the influence of the configuration and relative position of the cavity elements on the emission spectrum of a neodymium glass laser using a passive shutter. The laser cavity was fabricated so as to contain, other than the principal mirrors, no additional plane-parallel reflecting surfaces capable of affecting the spectrum of excited modes. Then the configuration and relative position of the individual cavity elements were varied and the changes in the emission spectrum investigated. In order to ascertain the role of the physical processes in the bleachable filter with respect to the formation of the emission spectrum we also investigated the spectrum of a neodymium glass laser in the free-oscillation regime and compared the results with the giant-pulse regime. Furthermore, so as to generalize the results to lasers with other active media and, in particular, to a medium with a homogeneous luminescence line width (as opposed to the inhomogeneous line width of neodymium glass) some of the investigations were carried out with ruby lasers.

The experimental laser consisted of a neodymium glass rod 220 mm long and 16 mm in diameter, with plane-parallel ends cut at the Brewster angle relative to the cavity axis,* and two external mirrors deposited on different types of substrates. Reflectors of dielectric coatings with a reflectivity of 0.5 to 0.6 at the wavelength $\lambda = 1.06\,\mu$ were deposited on a plane-parallel or wedge-shaped glass substrate to form the exit mirror R_2. The wedge angle ($\sim 5°$) was chosen so that the emitted radiation reflected from the second surface of the substrate would not revert to the active medium. The second mirror R_1 had a reflectivity $\geqslant 0.99$ and was also deposited either on a plane-parallel or a wedge-shaped substrate.† The cell containing the bleachable solution (initial transmissivity of the solution at $\lambda = 1.06\,\mu$: $T_0 \sim 0.7$) or the pure solvent was placed in the vicinity of R_1. The bleachable solution was a solution of No. 7 dye in nitrobenzene. The cell was filled with pure solvent in order to obtain comparable results in the free-oscillation and Q-switching regimes. The thickness of the cell was $d_k = 2$ mm, and its window thicknesses were 10 and 13 mm. The optical length of the cavity was $L = 90$ cm.

---

* This measure excluded the discriminative action of the rod ends on the emission spectrum and diminished the reflection losses.

† We also used mirrors deposited on a plane-parallel substrate with a matte finish on the second surface. They produced the same results as the mirrors with wedge-shaped substrates.

The emission spectra were studied for two different angles of inclination of the plane of the cell with respect to the cavity axis. In one case the angle of inclination α was equal to the Brewster angle α_B with an error no greater than 0.5°, and in the other case $\alpha \lesssim 1°$. In each case the laser emission spectrum was photographed in the two emitting regimes for different combinations of cavity mirrors. The emission spectrum was detected with a diffraction spectrograph having a linear dispersion $d\lambda/d \approx 2.5$ Å/mm and resolution ~0.1 Å.

Since a large number of spikes is emitted in the free-oscillation regime, even for a small excess ($\gtrsim 10\%$) above the emission threshold, for the investigation of the influence of the laser elements on the emission spectrum it is required to identify the resulting spectrum with each peak. This can only be done with time scanning of the spectrum. The same is true of the Q-switching regime when a train of giant pulses is emitted for a single pumping pulse. In order to circumvent this difficulty we worked in both regimes with only a slight excess above the emission threshold, such that only one free-oscillation peak or one giant pulse would be emitted in the period of one pumping pulse. Monitoring was realized by means of a photodiode connected to the input of an S1-16 oscilloscope.

The resulting spectrograms are illustrated in Figs. 23 and 24. For the case in which mirrors R_1 and R_2 on wedge-shaped substrates were used and the cell was oriented at the Brewster angle ($\alpha = \alpha_B$), when the laser cavity contained no parallel reflecting surfaces other

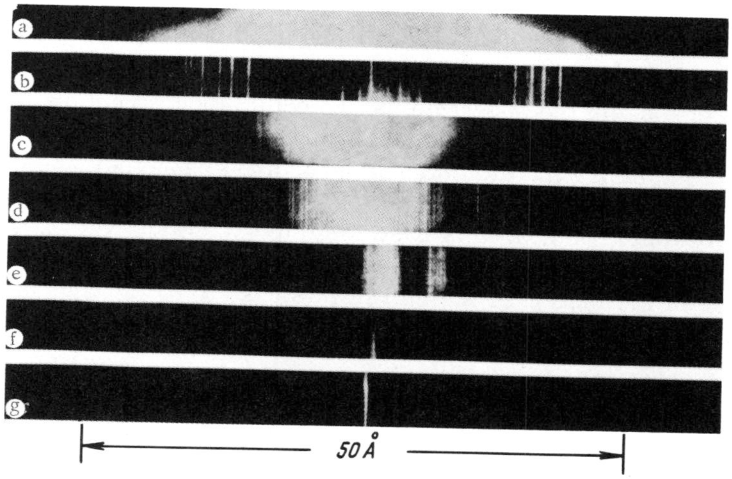

Fig. 23. Spectrograms of emission from a neodymium glass laser with a passive shutter (single giant pulse). a, b, c) Mirrors R_1 and R_2 on wedge-shaped substrates; c, d) mirror R_1 on plane-parallel substrate and R_2 on a wedge; e, f) R_1 and R_2 on plane-parallel substrates; a, c, e, g) $\alpha \approx \alpha_B$; b, d, f) $\alpha \lesssim 1°$; g) an identical cell, empty, placed near the cell containing the bleachable solution and inclined at an angle ~1°.

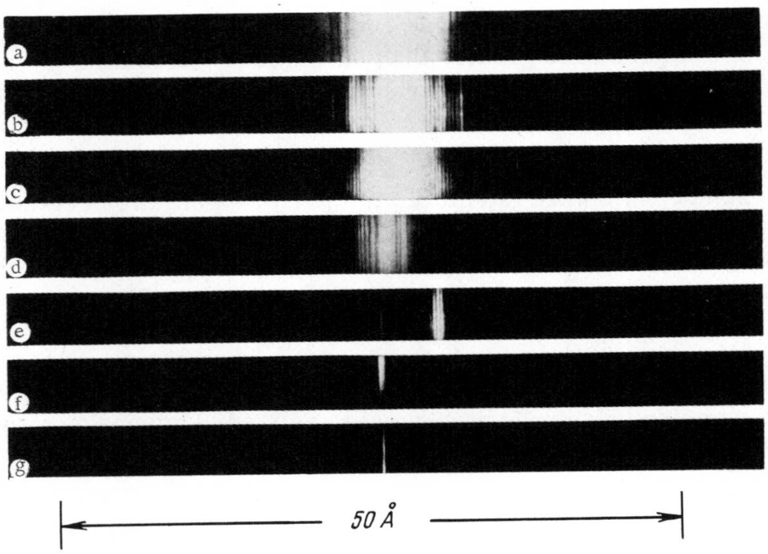

Fig. 24. Spectrograms of emission from a neodymium glass laser in free oscillation (single peak). The nomenclature is the same as in Fig. 23.

than R_1 and R_2, the emission has a continuous spectrum* (Figs. 23a and 24a). With the cell positioned at an angle $\alpha \leqslant 1°$ a rather complex, sharply defined structure is observed in the spectra (Figs. 23b and 24b), where it is induced by interference effects in the cell. This conclusion is supported by the dependence of the spectral structure on the linear dimensions of the structural details of the cell (thicknesses of the windows and the cell proper).

When the mirror R_1 on a wedge-shaped (or matte) substrate is replaced by a mirror on a plane-parallel substrate, again the spectrum has an equidistant structure for $\alpha = \alpha_B$ (Figs. 23c and 24d). The period of this structure coincides with the distance between the reflectivity maxima of the system comprising the dielectric mirror and second surface of the substrate: $\delta\nu_{str} = 1/2nd_{sub}$, where d_{sub} is the substrate thickness and n is the refractive index of the glass. Note that the reflectivity of mirror R_1 in this case is high, $\leqslant 0.99$, and yet the presence of the second parallel substrate surface with reflectivity ~ 0.04 is still felt in the resultant spectrum. A reduction of the angle of inclination of the cell in this case to $\alpha \leqslant 1°$ brings about a certain narrowing of the spectrum, as well as a refinement and sharper resolution of its structure (Figs. 23d and 24d).

With both mirrors R_1 and R_2 on plane-parallel substrates the structure of the spectra (Figs. 23e, 23f and 24e, 24f) is even more resolved, and the spectra themselves become still narrower (relative to the case depicted in Figs. 23c, 23d and 24c, 24d); this narrowing is more pronounced with the cell mounted at an angle $\alpha \leqslant 1°$ (Figs. 23f and 24f). The spectra in this last case degenerate into rather narrow lines of width $\Delta\lambda \leqslant 0.5$ Å.

Finally, if in addition to the cell containing the bleachable solution with $\alpha = \alpha_B$ another cell, empty, is inserted at $\alpha \leqslant 1°$ into the cavity with mirrors on wedge-shaped substrates,

*Actually the spectrum in this case as well consists of components which are equidistant modes of the principal cavity, but the spectral interval between them ($\delta\nu_m = 1/2L$) is too small to be resolved by our instrument. Evidence of the existence of modes is found in the beats observed in the signal intensity in this case, which have a frequency $f_1 = c/2L$.

where it functions as a complex Fabry–Perot etalon with four parallel surfaces, the spectra (Figs. 23g and 24g) exhibit only one line, whose width is determined by the resolution of the spectrograph: $\Delta\lambda \leqslant 0.1\,\text{Å}$. An investigation of this line by means of the Fabry–Perot interferometer in the Q-switching case showed that its width is $2\Delta\nu_{osc} \approx 0.07\ \text{cm}^{-1} \approx 13\delta\nu_m$, where $\delta\nu_m = 1/2L$. A similar result is obtained, as already stated in §2 of this chapter, when an active rod is used with end faces perpendicular to the cavity axis and the mirrors on plane-parallel substrates.

We have thus seen that by a suitable combination of selective elements in the cavity it is possible to vary the width of the emission spectrum of a neodymium glass laser with a passive shutter from $2\Delta\nu_{osc} \approx 30\ \text{cm}^{-1}$ to $2\Delta\nu_{osc} \approx 0.07\ \text{cm}^{-1}$ and even to $2\Delta\nu_{osc} < 3 \cdot 10^{-3}\ \text{cm}^{-1}$ (see §2 of this chapter). Since, as demonstrated by the experiments, the radiant energy in this case remains virtually unchanged, the foregoing result attests to the possibility of regulating the variation of the spectral density of the laser emission over a four-order-of-magnitude range.

A comparison of the emission spectra of neodymium glass lasers when internal mode discrimination is eliminated from the cavity (Figs. 23a and 24a) shows, first of all, that the width of the free-oscillation spectrum is much smaller (1/3 to 1/5) than the width of the giant-pulse spectrum. Also, there is an appreciable difference between the behavior of the wings of these spectra. Whereas the free-oscillation spectrum has a rather narrow frequency interval, $\Delta\nu = 20$ to $30\ \text{cm}^{-1}$, outside of which the spectral intensity is equal to zero even with the film highly overexposed (by a factor of 10^2 to 10^3), the wings of the giant-pulse spectrum occupy a frequency interval $\Delta\nu > 300\ \text{cm}^{-1}$. It is important to recognize, however, that the radiation intensity in these wings is much smaller (by 10^{-2} to 10^{-3}) than the intensity at the maximum. A detailed analysis of the possible causes of this disparity and an investigation of the envelope configurations of the spectra will be given below.

At this point we examine another difference between the free-oscillation and giant-pulse spectra in the case of passive Q switching. This difference is highlighted in Fig. 25, which shows the intensity distributions in the emission spectrum of the laser in both regimes (these distributions were obtained by photometric techniques). According to the figure, the spectral maxima have different positions on the wavelength scale: The giant-pulse maximum is shifted toward larger wavelengths by an amount $\delta\nu_s \approx 8.4\,\text{Å}$. This shift is induced by the fact that the maximum of the luminescence line of the active medium does not coincide with the maximum of the absorption band for the medium of the passive shutter (Fig. 26).

Proceeding from the foregoing assumption, we estimate the displacement of the giant-pulse spectrum relative to the free-oscillation spectrum (whose maximum in the given case is assumed to coincide with the maximum of the luminescence line for the active medium). The estimate we want can be obtained from the expression for the gain during the round-trip tran-

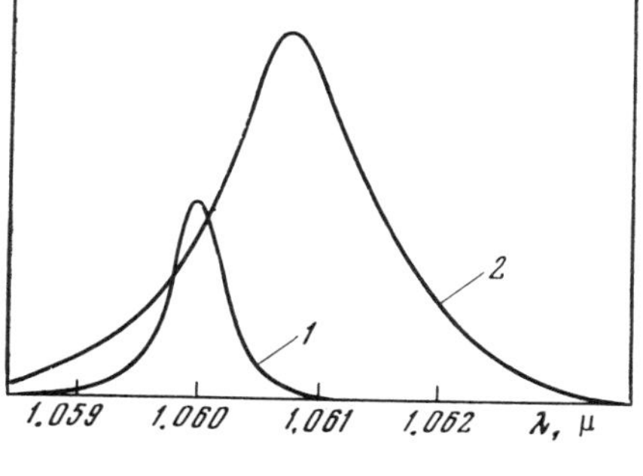

Fig. 25. Intensity distributions of the emission spectra of a neodymium glass laser (in relative units). 1) Single free-oscillation spike; 2) single giant pulse for passive switching.

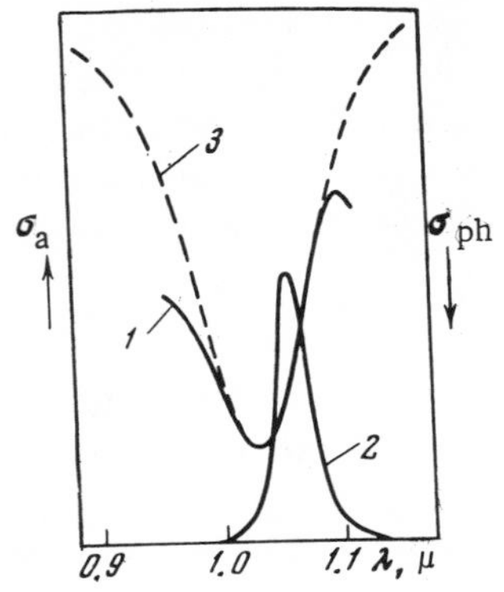

Fig. 26. Induced-transition cross sections versus wavelength. 1) Solution of No. 7 dye in nitrobenzene, σ_{ph}; 2) active medium, σ_a; 3) approximation of the dependence $\sigma_p(\lambda)$ by the dispersion curve 1.

sit time of a photon across the cavity: $G(\nu) = R_1R_2 \exp[-2\sigma_p(\nu) M_0 l_p] \exp[2\sigma_a(\nu) N l_a]$. Assuming a dispersion form for $\sigma_p(\nu)$ and $\sigma_a(\nu)$, we can represent $G(\nu)$ as follows:

$$G(\nu) = R_1R_2 \exp\left\{\frac{2\sigma_a N l_a}{1 + [(\nu - \nu_a)/\Delta\nu_a]^2} - \frac{2\sigma_p M_0 l_p}{1 + [(\nu - \nu_p)/\Delta\nu_p]^2}\right\}, \quad (15)$$

where ν_a and ν_p are the values of the frequencies at which the maxima of the luminescence line for the active medium and the absorption band for the filter occur and $2\Delta\nu_a$ and $2\Delta\nu_p$ are the widths of the luminescence band for the active medium and the absorption band for the filter medium, respectively. The quantities $2\sigma_a N l_a$ and $2\sigma_p M_0 l_p$ are determined from the conditions $G = 1$ and $T_p^2 = \exp(-2\sigma_p M_0 l_p)$, where T_p is the initial transmissivity of the passive shutter at the frequency ν_p. To find the frequency ν_{max} with the maximum gain $G(\nu_{max})$ we need to determine the roots of the equation $dG/dx = 0$, where $x = (\nu - \nu_a)/\Delta\nu_a$. However, the fifth-degree equation in x obtained in this case has, in general, five roots, so that we apply an approximative method here, because for the majority of dyes that we use $|x| < 0.1$. Thus, in the case of a neodymium glass laser with the following system parameters: $R_1 \approx 1.0$, $R_2 = 0.55$, $T_0 = 0.6$, $T_p = 0.5$, $\nu_p - \nu_a \approx 250$ cm^{-1}, and $2\Delta\nu_p \approx 1000$ cm^{-1} the approximate calculation yields $(\nu_{max} - \nu_a) \cdot 1/\Delta\nu_a \approx 0.066$, the minus sign denoting a shift toward lower frequencies. For $2\Delta\nu_a = 250$ cm^{-1} the shift of the giant-pulse emission maximum relative to the free-oscillation maximum should be $\nu_{max} - \nu_a \approx -8.3$ cm^{-1}. According to Fig. 25, the experimental value $(\nu_{max} - \nu_a)_{exp} \approx -7.5$ cm^{-1}, which is in good agreement with the calculated value.

We note that the selective properties of a plane-parallel layer with respect to the Q-values of certain modes of the cavity are retained with oblique incidence of a plane wave, because in this case as well the transmissivity of the layer depends on the wavelength. However, for a given layer thickness t and a given aperture d of the parallel light beam there is a limiting angle of inclination α_{lim} above which the beam, on being doubly reflected from the surfaces of the layer, does not intersect the straight-through beam (Fig. 27). In this case interference effects should no longer occur. An elementary calculation gives the following relation between the quantities t, d, and α_{lim}:

$$\frac{t \sin 2\alpha_{lim}}{\sqrt{n^2 - \sin^2\alpha_{lim}}} = d, \quad (16)$$

in which n is the refractive index of the layer material.

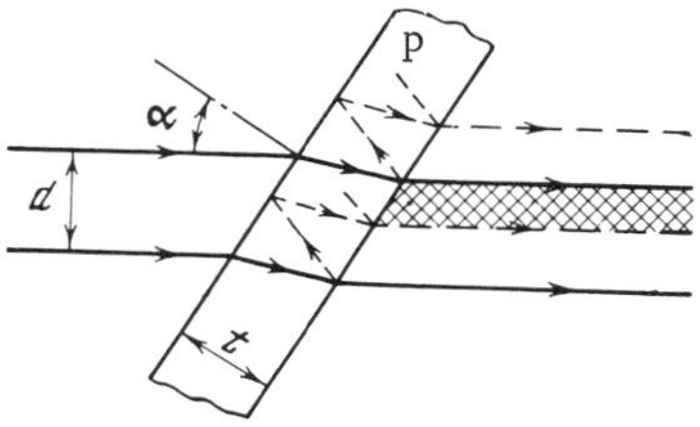

Fig. 27. Propagation of a plane electromagnetic wave of bounded aperture angle across a plane-parallel plate. Interference occurs only in the cross-hatched region.

The following experiment was set up to confirm the validity of this equation and to determine the limiting inclination of the plane-parallel plate. Inside a cavity with mirrors R_1 and R_2 on wedge-shaped substrates, separated by a distance L = 300 cm, we placed a plane-parallel glass plate of thickness t = 12.8 mm (n = 1.65) mounted on a goniometer stage. We were able to measure the plate angle α in steps of 10^{-3} rad ≈ 3.5'. Diaphragms 4.2 mm in diameter were inserted in the cavity near the ends of the active rod, where they restricted the aperture of the emission beam and its divergence associated with the excitation of high-order transverse modes. Under these conditions only TEM_{00p} modes were excited in the cavity. We then photographed a series of spectrograms for various angles α. The fine structure disappeared from the spectrograms for $\alpha > \alpha_{lim} = 0.318 \pm 0.007$ rad. The substitution of this value into Eq. (16) gives an aperture value d = 4.7 ± 0.1 mm, which slightly exceeds the diameter of the diaphragm. The difference is clearly attributable to the finite divergence of the laser emission due to diffraction at the diaphragm, so that the intensity of the field beyond the limits of the diaphragm does not fall off instantaneously. Equation (16), however, was derived for the case of a sharply confined beam aperture.

The results of this section indicate that the selective attributes of various cavity elements exert a powerful influence on the emission spectrum of a neodymium glass laser. This selection induces approximately identical changes in the structure of the spectrum in both the free-oscillation (single spike) and the passive Q-switching regime with a bleachable filter, thus evincing a certain commonality between the pulse-buildup mechanisms in both regimes. An important practical inference drawn from our results is the following. In order to obtain a "continuous" emission spectrum of the largest possible width for a neodymium glass laser, the possible selective action of various cavity elements must be eliminated from the cavity, i.e., all elements present with plane-parallel surfaces must be oriented either at the Brewster angle or at an angle $\alpha > \alpha_{lim}$. Orientation at the Brewster angle, however, is preferable, because parasitic reflection losses are absent in this case, and the laser emission is plane-parallel.

In order to generalize the latter result, we investigated the spectrum of a ruby laser with a passive shutter and with the elimination of selectivity from the cavity [23]. The active rod was ruby, 15 mm in diameter and 120 mm long, with plane-parallel end faces cut at the Brewster angle. The cavity mirrors, deposited on wedge-shaped backings (wedge angle ~ 3.6°), had reflectivities $R_1 = 0.86$ and $R_2 = 0.55$. A cell containing a solution of cryptocyanine in nitrobenzene (passive shutter) was placed near R_1 and tilted at the Brewster angle with respect to the cavity axis. The initial transmissivity of the cell was $T_0 = 0.38$. The optical length of the cavity was L = 85 cm. A diaphragm 4 mm in diameter was inserted in the cavity. The angle of divergence of the emission in this case was four or five times the angle of divergence of the TEM_{00p} modes. The emission spectrum was investigated with a Fabry–Perot interferometer having a plate separation $d_{FP} = 1$ mm and a resolution no worse than 0.1 cm^{-1}. Under these conditions the spectrum of a single giant pulse had a total width $\Delta\nu \approx 3$ cm^{-1} (Fig. 28). We note that this is the first time a giant pulse with such a broad spectrum has been obtained in a ruby laser using a passive shutter. For comparison we cite the similar study [19], in which

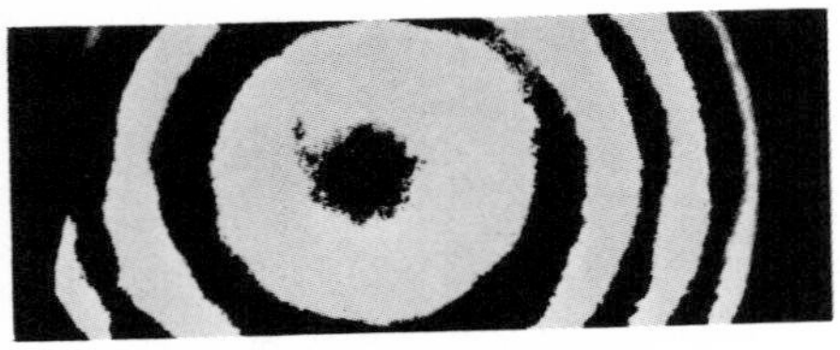

Fig. 28. Interference pattern of the emission spectrum of a ruby laser with a passive shutter (solution of cryptocyanine in nitrobenzene) in the single giant-pulse regime. Spacing of interferometer plates: $d_{FP} = 1$ mm.

the width of the laser emission spectrum was about 0.03 cm^{-1}. We did not observe any fine structure in our spectrum. Consequently, it may be concluded on the basis of the given experiment that the specific aspects of the broadening of the amplification (or spontaneous emission) line of the active medium is of secondary value in the formation of the stimulated emission spectrum for a giant pulse (passive switching) and (single) free-oscillation spike, the selective properties of the cavity being of prime importance in this respect.

CHAPTER III

DEVELOPMENT OF THE EMISSION PULSE IN THE CASE OF PASSIVE SWITCHING

§1. Giant-Pulse Buildup in a Q-Switched Laser

We mentioned in Chapter I that certain characteristics of a giant pulse emitted by a laser with a passive shutter are well accounted for on the basis of the theory of lasers with an instantaneous shutter. The passive shutter, however, also has its own specific peculiarities, which distinguish it from other types of shutters. In this chapter we consider one of its more remarkable properties, which, as stated earlier, makes it possible to vary the width of the stimulated emission spectrum through several orders of magnitude by rather simplistic structural modifications of the cavity of lasers using such a shutter. In the case of a neodymium glass laser this variation amounted to about four orders of magnitude, ranging from $2\Delta\nu_{osc} \approx 30$ cm^{-1} to $2\Delta\nu_{osc} < 3\cdot10^{-3}$ cm^{-1} [22, 82]. This property is caused by a certain singularity in the formation of the giant pulse in a laser with a passive shutter, being attributable to the fact that the formation period, measured from the time at which the gain in the active medium offsets the cavity losses to the instant at which the pulse intensity attains its peak value, is considerably larger than for other types of shutters.

The giant-pulse buildup time in a laser with a passive shutter was first estimated theoretically in [83]. In the same paper an explanation is given for the appreciable variations of the spectral composition for lasers with a passive shutter by comparison with other types of shutters, on the basis of the given estimate and certain assumptions with regard to the formation mechanism. In this study the process of formation of a giant pulse from spontaneous noise is treated as an independent growth of the radiation intensity in the individual modes. The entire buildup process is divided into short time intervals, $t_c = 2L/c$, equal to the round-trip transit time of a photon over the length L of the cavity. In the course of this interval the total gain in the cavity is assumed to be constant and equal to the following for the n-th mode:

$$G_n = R_1R_2Q_n \exp\,[2\sigma_{an}N(t)\,l_a], \tag{17}$$

where Q_n represents the losses inside the cavity and N(t) is the population inversion.

The power of the mode at time $t_q = qt_c$ in this case is expressed as

$$P_n(t_q) = P_n(0)\prod_{i=1}^{q} G_{ni}, \tag{18}$$

where $P_n(0)$ is the power of the n-th mode at the beginning of oscillation and G_{ni} is the single-pass gain at time t_i. For the case of instantaneous switching (ideal shutter) we can put N(t) = const and Q_n = const. Then $G_{ni} = G_n$ = const, and $P_n(t_q) = P_n(0)\,G_n^q$. The number q of passes required for the mode power to grow from the spontaneous noise level [$P_n(0) \sim 10^7$ photons/sec · mode] to the typical giant-pulse power level [$(P_n)_{max} \sim 10^{24}$ photons/sec · mode] is very easily estimated in this case. Setting $G_n \approx 10^{1/2}$, we obtain $G_n^q = 10^{q/2} = (P_n)_{max}/P_n(0) \approx 10^{17}$, or $q \approx 34$, i.e., the time required for the buildup of a giant pulse in a laser incorporating an ideal shutter is about $34t_c$. If for some reason (say, $Q_n/Q_m \neq 1$) the gain is not identical for two modes, the power difference between these modes will increase with q. In order for the power ratio of two modes at the time of the pulse peak to be equal to $P_n(t_q)/P_m(t_q) = 10$ for q = 34 it is sufficient for the initial relative difference between the losses of these modes to be about 7%, i.e., $Q_n/Q_m = 1.07$. This fact accounts for the narrowing of the giant-pulse emission spectrum relative to the width of the amplification line in the active medium.

In the case of a laser with a passive shutter the pulse buildup process can be conditionally divided into two intervals. On the first interval the gain G_n can be assumed close to unity, but dependent on the time, because N = N(t). The cavity losses Q_n are determined mainly by the losses in the passive shutter, i.e., $Q_n = T_{0n}^2$, and are assumed to be constant over the entire first interval. Inasmuch as $G \approx 1$, the power growth on this interval is slow. When a certain threshold power level is reached ($P_{thr} \sim 10^{22}$ photons/sec · mode), the continued power growth (second interval) will be accompanied by a decrease of the filter transmissivity T_0. This causes a rapid decrease in the population inversion. This interval (which may be called the nonlinear interval, since the transmissivity of the shutter depends nonlinearly on the radiation power) is usually of much shorter duration, so that it suffices to determine the length of the first, linear interval in order to estimate the giant-pulse buildup time for a laser with a passive shutter.

For the estimation of this time in [83] the dependence N(t) was assumed to be linear: $N(t) = \eta t$. Now the oscillation inception time for the n-th mode [$G_n(t) = 1$] is expressed as

$$t_n = -\frac{\ln(R_{1n}R_{2n}T_{0n}^2)}{2\sigma_{an}\eta l_a}, \tag{19}$$

and the intensity variation of the n-th mode in time t_c is expressed as

$$\frac{dP_n(t)}{dt} \approx \frac{P_n(t)}{t_c}\{\exp[2\sigma_{an}\eta l_a(t-t_n)]-1\}. \tag{20}$$

The following expression is readily obtained for the intensity of this mode at the time t_s of initiation of bleaching of the filter:

$$P_n(t_s) = P_n(t_n)\exp\left[\frac{\sigma_{an}\eta l_a(t_s-t_n)^2}{t_c}\right]. \tag{21}$$

Inasmuch as the required increase in intensity must be of the order $P_{thr}/P_0 \approx 10^{15} \approx e^{35}$, from the latter expression we obtain

$$\frac{\sigma_{an}\eta l_a(t_s-t_n)^2}{t_c} = 35. \tag{22}$$

Setting $\eta = 1.5 \cdot 10^{23}$ sec^{-1}, $l_a = 7.5$ cm, $\sigma_{an} = 2 \cdot 10^{-20}$ cm^2, and $t_c = 3 \cdot 10^{-9}$ sec, the author of [83] obtained a value of ~ 720 for the quantity $q = (t_s - t_n)/t_c$; this value considerably exceeds the value of q for the case of instantaneous Q switching. This model therefore well accounts for the so-called selective property of a passive shutter, i.e., the intensification of the difference between the Q-values of the modes due to the selective action of the cavity elements.

A later, more rigorous calculation [64] of the processes involved in lasers using a passive shutter bore out the validity of the fundamental results of the simple model described above. This calculation was based on the set of equations (9), which was solved on a computer for certain predetermined parameters. In particular, the time dependence of the photon density in the laser cavity, according to the calculation, appears as shown in Fig. 29. As this graph indicates, over its principal interval of variation ($0 < t < 200t_{ph}$) the growth of the photon density in the cavity obeys the law $\Phi = \Phi_0 \exp[\text{const}\,(t/t_{ph})^2]$, as in the model of [83]. Only in the very peak region (immediate vicinity of the giant pulse) does the variation of $\Phi(t)$ jump into a more abrupt type of law. At this time the so-called filter bleaching process begins, i.e., the transmissivity of the filter changes under the action of the incident radiation. At the same time, a fundamental change is experienced by such parameters as the active-particle population inversion N, photon density Φ, and population difference M for the passive filter particles. The time variation of these parameters in the immediate interval of the giant pulse is illustrated in Fig. 30.

An experimental estimate of the giant-pulse buildup time in a laser with a passive shutter was first obtained in [88]. In this study a ruby laser with a passive shutter (solution of vanadium phthalocyanine in nitrobenzene) was used, operating in the single-mode regime (one TEM_{00p} mode), so that a very high reproducibility was obtained with respect to the parameters of the giant pulse. This fact enabled the authors to investigate the dependence of the output power

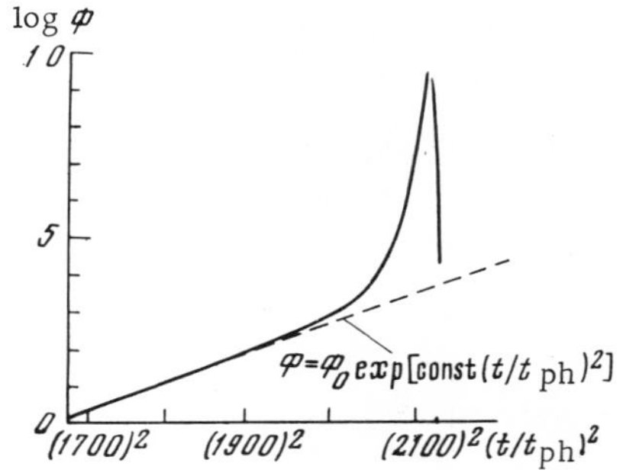

Fig. 29. Photon density Φ (in relative units) in the cavity versus time for a laser with a passive shutter (after [64]).

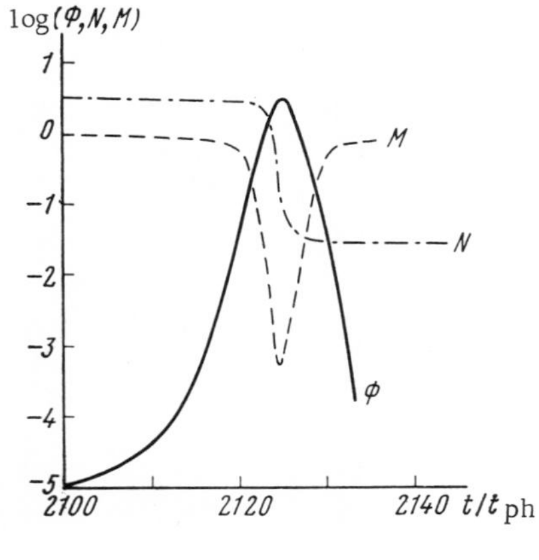

Fig. 30. Time variation of cavity photon density Φ, population inversion N, and population difference M for the filter particles in a laser with a passive shutter (after [64]).

P_{out} (which is proportional to the cavity photon density Φ) on the distance (measured in units of t_c) to the pulse peak, with the range of variation of the power spanning four orders of magnitude. The resulting curve, plotted in coordinates $(t/t_c)^2$ and ln P_{out}, exhibited for low power (or Φ) levels a linear interval, which was then approximated down to the spontaneous noise level P_0. In this approximation a value of $\sim 1150t_c$ was obtained for the pulse buildup time in the ruby laser with a passive shutter.

§2. Determination of the Giant-Pulse Buildup Time in a Neodymium Glass Laser with a Passive Shutter

In measurements of the giant-pulse buildup time in lasers using a passive shutter by the method proposed in [88] exceedingly high demands are imposed on the output characteristics of the laser from the standpoint of constancy of the laser parameters from one pulse to the next. These demands, obviously, can only be met in single-mode operation, which is by no means always realizable.

In [89] we proposed another, simpler in practice, method for estimating the buildup time of a giant pulse in a laser with a passive shutter and used it to determine the pulse buildup time in a neodymium glass laser. The method is based on a measurement of the emission intensity ratio of two different modes (or groups of modes) at the time of the giant-pulse peak and a comparison of the value so obtained with the Q ratio of these modes at the beginning of the pulse buildup period. It is assumed (in accordance with [64]) that the buildup time is determined mainly by the growth time of the photon density in the cavity from its initial level to the value at which nonlinear absorption begins in the shutter. On this interval the absorption in the shutter can be regarded as constant and equal to the initial value. Inasmuch as the pulse buildup interval, on which the shutter absorption is nonlinear, is considerably shorter than the first interval, it may be assumed that no appreciable redistribution of intensity between widely spaced modes will take place in this time. It is essential to realize that the nonlinear absorption process in the passive shutter can be accompanied by self-locking of the individual modes with respect to phase [90] and, clearly, a certain equalization of their amplitudes. This effect will tend somewhat to smooth out the mode intensity distribution formed at the beginning of nonlinear absorption in the shutter, so that our estimate of the giant-pulse buildup time represents a lower limit of its true value.

Under the given assumptions we regard the temporal development of the giant-pulse spectrum as an independent growth of the intensities of the individual modes with the following gain for the n-th mode in the time t_c: $G_n(t) = R_1R_2T_{0n}^2 \exp[2\sigma_{an} N(t) l_a]$. For closely spaced modes n and m we can put $\sigma_{an} = \sigma_{am} = \sigma_a$ and $T_{0n} = T_{0m} = T_0$. Moreover, we shall assume that $R_{2n} = R_{2m} = R_2$ and $|\Delta R| = |R_{1m} - R_{1n}| \ll R_{1n}$. Since the pulse buildup time is considerably shorter than the pumping time or the spontaneous decay time of the upper working level of the active medium, the dependence N(t) may be assumed to be linear: $N = \eta t$, where $\eta > 0$. Then, according to [83], the intensity ratio of the two modes at time t_s (onset of nonlinear absorption in the shutter) is expressed as

$$\frac{P_m(t_s)}{P_n(t_s)} = \frac{\exp\left[\frac{1}{t_p}\sigma_a\eta l_a (t-t_m)^2\right]}{\exp\left[\frac{1}{t_p}\sigma_a\eta l_a (t-t_n)^2\right]}, \qquad (23)$$

where t_n and t_m are the times of emission inception for the corresponding modes. The expressions for t_n and t_m have the following form due to the smallness of ΔR:

$$t_n = -\frac{\ln(R_{1n}R_2T_0^2)}{2\eta\sigma_a l_a}, \qquad t_m \approx t_n - \frac{\Delta R}{2\eta\sigma_a l_a R_{1n}}. \qquad (24)$$

Assuming $t_s - t_n \gg |t_n - t_m|$, we readily simplify expression (23):

$$\frac{P_m(t_s)}{P_n(t_s)} = \exp\left\{\frac{\sigma_a \eta l_a}{t_p}\left[2(t_s - t_n)(t_n - t_m) + (t_n - t_m)^2\right]\right\} \approx \left[\exp\left(\frac{\Delta R}{R_{1n}}\right)\right]^{\frac{t_s - t_n}{t_p}} \approx \left(1 + \frac{\Delta R}{R_{1n}}\right)^q, \tag{25}$$

where $q = (t_s - t_n)/t_c$ is the giant-pulse buildup time (to the onset of nonlinear absorption) in units of t_c.

It is apparent from Eq. (25) that if the difference ΔR of the reflectivities is known for two modes (or groups of modes) and the intensity ratio of these modes in the emission spectrum of a single giant pulse has been measured, it is a simple matter to determine the quantity we are after, viz., q. Inasmuch as we used photometric techniques to measure the relative intensity of different groups of modes, only for $P_m(t_s)/P_n(t_s) \leqslant 10$ (interval of linear dependence of the film optical density on the log intensity for the particular type of film used) were we able to obtain a sufficiently accurate measurement of this quantity. According to (25), the required variation of the reflectivity for $q \sim 10^3$ must then be of the order $|\Delta R|/R_m \leqslant 2.3 \cdot 10^{-3}$. We achieved this small (measured) variation of the reflectivity for separate mode groups by depositing a mirror with reflectivity $\geqslant 0.99$ on a plane-parallel substrate. As we showed in [22], the presence of Fresnel reflection from the opposite side of the substrate leads (for an appropriate cavity design) to an appreciable intensity variation of different parts of the spectrum in connection with interference effects at this mirror.

The maximum variation of the reflectivity (with respect to intensity) $\Delta R = R_{max} - R_{min}$ for such a complex mirror with interference effects taken into account is easily calculated. According to [49], the amplitude reflectivity of the system illustrated in Fig. 31, which comprises two parallel reflecting surfaces, is equal to

$$r(\lambda) = \frac{r_{12} + (\delta_{12}\delta_{21} - r_{12}r_{21})\, r_{23} \exp\left(-i\,\frac{4\pi n_2 t}{\lambda}\right)}{1 - r_{21}r_{23}\exp\left(-i\,\frac{4\pi n_2 t}{\lambda}\right)} = \frac{-r_1 + (1 - a^2)\, r_2 \exp\left(-i\,\frac{4\pi n t}{\lambda}\right)}{1 - r_1 r_2 \exp\left(-i\,\frac{4\pi n t}{\lambda}\right)}, \tag{26}$$

where r_{ij}, δ_{ij}, and a are the amplitude reflectivity, transmissivity, and absorptivity at the corresponding boundary with regard for the direction of wave propagation. It is evident from (26) that $r(\lambda)$ is a periodic function of the wavelength and has the following maximum and minimum values:

$$|r|_{max} = \frac{r_1 + (1 - a^2)\, r_2}{1 + r_1 r_2}, \qquad |r|_{min} = \frac{r_1 - (1 - a^2)\, r_2}{1 - r_1 r_2}. \tag{27}$$

Inasmuch as in our case $|\Delta r| = |r|_{max} - |r|_{min} \ll |r|_{max} \approx 1$, our required variation ΔR is ex-

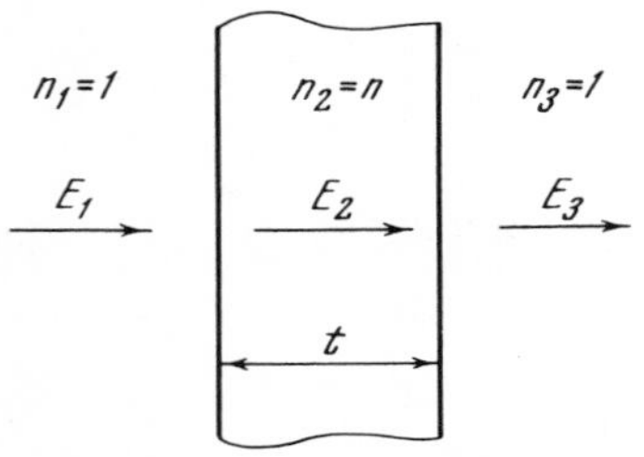

Fig. 31. System of two reflecting layers. $\delta_{12}\delta_{21} - r_{12}r_{21} = 1 - a^2$, where a is the amplitude reflectivity at the boundary 1-2; $r_{21} = -r_{12} = r_1 > 0.99$; $r_{23} = r_2 = 0.2$; $\delta_{23} = 2n/(n+1)$; $\delta_{32} = 2/(n+1)$.

pressed as follows:

$$\Delta R = |r|^2_{\max} - |r|^2_{\min} \approx 2\Delta r = \frac{2r_2(1 - a^2 - r_1^2)}{1 - r_1^2 r_2^2} = \frac{2r_2\delta_{12}\delta_{21}}{1 - r_1^2 r_2^2}. \quad (28)$$

As Eq. (28) indicates, for $r_1 \approx 1$ the variation ΔR depends in the final analysis only on the reflectivity r_2 of the second boundary and the transmissivity $\delta_{12}\delta_{21}$ of the first boundary, and not on $(1 - r_1^2)$. This fact greatly facilitates the obtaining of controllable small variations ΔR, because it is easier by far in practice to measure small transmissivities than to measure large reflectivities r_1.

The apparatus used in this experiment comprised an ordinary plane-parallel laser cavity, in which special care was exercised to eliminate the mode discrimination effect due to various elements, except in the case of the mirror R_1, which was deposited on a plane-parallel glass substrate. The exit mirror R_2, with a reflectivity of ~ 0.7, was deposited on a wedge-shaped substrate with vertex angle $\sim 5°$. The active medium was a KGSS-7 glass rod 15 mm in diameter and 220 mm long, with end faces cut at the Brewster angle relative to the cavity axis. The passive shutter was a cell containing a solution of No. 7 dye in nitrobenzene, tilted at the Brewster angle relative to the cavity axis. The initial transmissivity of the cell was 0.65. The single-pulse regime was monitored with a photodiode connected to an S1-16 oscilloscope. The spectral composition of the emission was investigated with a diffraction spectrograph having a linear dispersion $d\lambda/d \approx 2.5$ Å/mm and resolution of ~ 0.1 Å. The thickness of the substrate for R_1 ($d_{sub} = 1.5$ mm) was chosen so that the distance between its reflection maxima ($\delta\nu_{sub} = 1/2nd_{sub} \approx 2.3$ cm^{-1}) would be much greater than the width of the spectrograph instrument function and smaller than the width of the emission spectrum.

In order to determine the factor $\delta_{12}\delta_{21}$ in Eq. (28) we measured the transmissivity (with respect to intensity) of R_1 on a spectrometer; the spectral width of the slit was much greater than the distance between the transmission interference maxima of the substrate, hence these effects could be neglected. In this case the transmissivities of the substrate in both directions are equal: $T_{13} = \delta_{12}^2\delta_{23}^2 = \delta_{32}^2\delta_{21}^2 = T_{31}$. Inasmuch as the coefficients δ_{23} and δ_{32} can be determined once the refractive index n of the substrate is known, neither does the determination of δ_{12} and δ_{21} present any special difficulty. In our case the product of these proved to be equal to $\delta_{12}\delta_{21} \approx 2 \cdot 10^{-3}$, so that the maximum variation of the reflectivity of R_1 was $\Delta R \approx 1.6 \cdot 10^{-3}$.

The contrast of the fine structure of the laser emission spectrum (Fig. 32) was measured by the usual photographic photometric techniques. In order to obtain the optical density curve a density reference mark was superimposed on the film with the investigated spectra by photographing the image of the spectrograph exit slit (which was illuminated with the laser radiation by means of a spherocylindrical condenser) in the zeroth order of a diffraction grating through neutral filters with the measured transmissivity. The laser in this case operated in the free-oscillation regime, and the relative variation of the radiant energy from one burst to another did not exceed 10%.

A measurement of the emission intensity ratio for adjacent maxima and minima of the resulting spectrograms showed that this ratio is a maximum at the center of the spectrum and

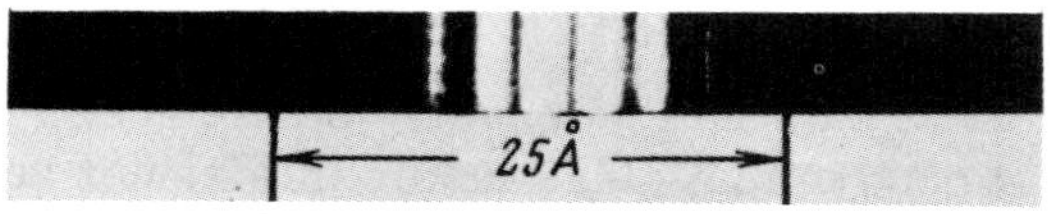

Fig. 32. Spectrogram of a giant pulse emitted by a neodymium glass laser with a passive shutter (plane-parallel mirror R_1).

decreases to 1.0 toward the edges of the spectrum. The scatter of this ratio (at the center) for several spectrograms ranged from 2.5 to 3.1. Since our estimate represents a lower estimate of the buildup time, we let $P_{max}/P_{min} = 3.1$ and for $\Delta R = 1.6 \cdot 10^{-3}$ obtain, according to (25), $q > 700$ or $t_s - t_n > 4$ μsec.

Consequently, the value obtained for the number q of passes required for buildup of a giant pulse in a neodymium glass laser with a passive shutter, given the assumption that the nonlinear interval of the pulse transient has little effect on the resultant spectrum, turned out to be of the same order as the value of q measured for a single-mode ruby laser [88]. The value obtained for q also affords strong evidence of the considerable influence of the initial buildup interval of the giant pulse on the resultant emission spectrum of the laser with a passive shutter.

§3. Determination of the Buildup Time for a Free-Oscillation Spike in a Neodymium Glass Laser

The method we proposed in [89] for estimating the emission pulse buildup time, having been successfully used for a neodymium glass laser with a passive shutter, can also be used in principle for the free-oscillation regime if the latter meets certain conditions. First of all, it is required that the initial pulse interval, in which the main formation of the emission spectrum takes place, be of much greater duration than the final interval, in which the main radiation of pulse energy occurs due to the stored population inversion. This requirement is equivalent to stating that the variation of the population inversion and photon density in the cavity must proceed much more slowly on the initial interval than on the final interval. Also, the variation of all parameters on the first interval must be slow enough to be considered constant on the interval t_c. Then the entire pulse buildup process can be separated into distinct time intervals t_c, on which the gain for the n-th mode can be assumed constant and equal to $G_n = R_{1n}R_{2n}Q_n \exp[2\sigma_{an}N(t)l_a]$. By the proper choice of cavity design a situation can be created in which only R_1 is left depending on n, whereupon the intensity ratio of two modes at time $t_q = qt_c$ will be expressed as follows for $|R_{1m} - R_{1n}| = |\Delta R| \ll R_{1n}$:

$$\frac{P_m(t_q)}{P_n(t_q)} = \left(\frac{R_{1m}}{R_{1n}}\right)^q \approx \left(1 + \frac{\Delta R}{R_{1n}}\right)^q. \tag{29}$$

All of the foregoing requirements are satisfied, in particular, in the emission of a single free-oscillation spike. Above all, if the pumping energy is not too far above the threshold level, the initial buildup phase of the spike will be several times the width of the spike itself [91].

The total buildup time of the free-oscillation spike (being of the same order of magnitude as the distance between adjacent spikes in the free-oscillation regime with several spikes) is much greater than the transit time t_c. Consequently, our method for estimating q is clearly applicable to this case as well.

The experimental arrangement was the same as before, except that the passive shutter was excluded from the laser cavity. The laser operated with the emission of a single free-oscillation spike, this regime being achieved by making the pumping energy just slightly in excess of the threshold level. The variation $\Delta R \approx 1.6 \cdot 10^{-3}$ was obtained by having the mirror R_1 on a plane-parallel substrate. A measurement of the structural contrast of the resulting spectrograms (Fig. 33) (by the photographic photometric technique) showed that the ratio $P_m(t_q)/P_n(t_q)$ varies between the limits from 8 to 12, corresponding to values of q = 1300 to 1600.

Thus, the value of q is of the same order for a giant pulse in the passive shutter case and for the emission of a single free-oscillation spike in a neodymium glass laser, this result being

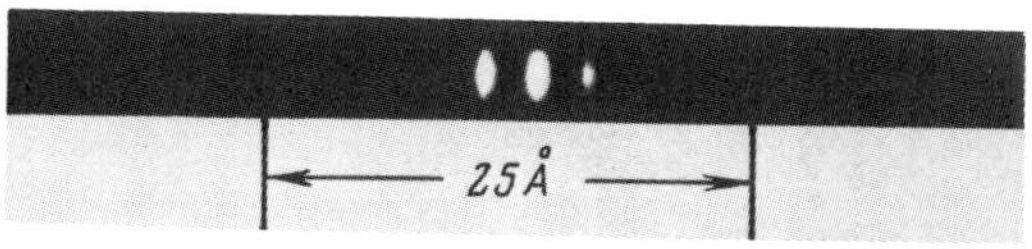

Fig. 33. Spectrogram of the emission of a single free-oscillation spike from a neodymium glass laser (plane-parallel mirror R_1).

attributable to the qualitative similarity of the behavior of the emission spectrum in the two regimes with the insertion of various discriminating elements into the cavity [22].

CHAPTER IV

SELF MODE LOCKING IN A SOLID-STATE LASER USING A BLEACHABLE FILTER

§1. Theoretical Model of the Self-Locking Process

As we mentioned in the Introduction, the locking of axial modes can be realized independently in lasers incorporating a bleachable filter, without external modulation. This effect, which is known as the self-locking of modes, was first observed in our investigation of the time characteristics of the emission from a neodymium glass laser with a passive shutter [18], as well as in [19, 20]. In the case of self-locking the giant pulse had the form (Fig. 34) of a train of separate shorter pulses with a fairly large spacing, the latter usually being equal to 2L/c.

The first theoretical account of the self-locking process in a laser with a bleachable filter was offered by Kuznetsova in our joint paper [90].* The following model was considered in the paper. The steady state was assumed, i.e., the filter bleaching and field buildup processes were disregarded. The amplitudes of the various spectral components of the field were considered to be constant and given. Under these assumptions the absorption of energy of the field in the filter was determined for arbitrary phases of the spectral components, and the phase relations were found for which the energy losses are minimal. It is supposed that the operating regime in which these relations are satisfied is the one that will be realized in the laser. The medium of the filter in this case is described by the two-level scheme

$$\begin{aligned} &\frac{d}{dt}(\rho_{11}-\rho_{22})+\gamma(\rho_{11}-\rho_{22})=\gamma N-2i\frac{p}{\hbar}E(\rho_{12}-\rho_{12}^{*}),\\ &\frac{d}{dt}\rho_{12}+(\Gamma-i\omega_0)\rho_{12}=-i\frac{p}{\hbar}E(\rho_{11}-\rho_{22}). \end{aligned} \tag{30}$$

Here ρ_{ij} are the elements of the density matrix; γ and Γ are the relaxation constants of the populations and off-diagonal element, where $\Gamma \gg \gamma$; N is the population difference between the lower and upper levels in the absence of the field; ω_0 is the transition eigenfrequency; p is the dipole moment matrix element; and E is the electric field strength.

The field is assumed to consist of several discrete spectral components (modes), the spatial distribution for each component being close to a standing plane wave. In this case it

* A theoretical model of self-locking in a laser with a bleachable filter has also been considered independently in [92]. The authors of this paper, however, having limited the number of modes to three, were unable to deduce certain laws manifested only in the excitation of many (more than three) modes.

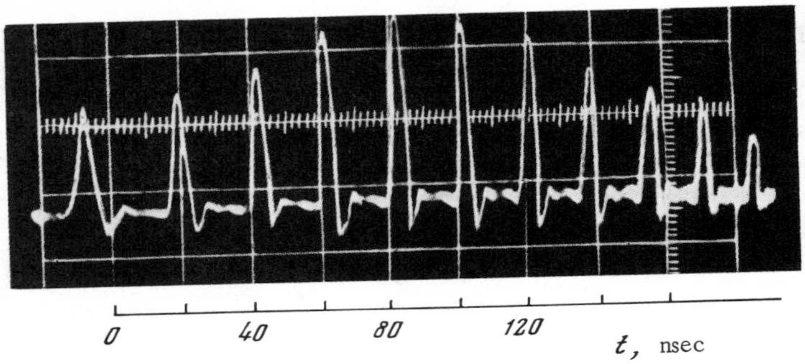

Fig. 34. Oscillogram of a giant pulse from a neodymium glass laser with a passive shutter (No. 7 dye in nitrobenzene; $T_0 = 71\%$). L = 282 cm; distance from mirror R_1 to cell: $X_1 = 2$ cm; $L_1 = 94$ cm; sample tilted ~ 7' relative to cavity axis.

may be represented as

$$E = \sum_m A_m \sin k_m X \cos(\omega_m t + \varphi_m) \equiv \frac{\hbar}{p} \sum_m F_m \cos(\omega_m t + \varphi_m), \tag{31}$$

where $A_m > 0$, $k_m = \omega_m / c$, $\omega_m = \omega + m\Omega$, and $\Omega = \pi c / L$.

Next the electromagnetic energy W absorbed by the filter medium per unit time is calculated. On the assumption that the polarization P of the filter medium is associated entirely with the given pair of levels:

$$P = np(\rho_{12} + \rho_{12}^*), \tag{32}$$

where n is the density of absorbing centers, the following expression is obtained for the average electromagnetic energy over the period $2\pi/\omega$:

$$W_{av} = \left(E \frac{dP}{dt}\right)_{av} = n\hbar\omega \frac{N}{\Gamma} \left\{ \left(\sum_r F_r^2\right)\left(1 + \Gamma^{-1}\gamma^{-1} \sum_m F_m^2\right)^{-1} - \frac{\gamma}{2\Gamma} \sum_s \sum_l \sum_m F_m F_l F_s F_{m+s-l} \frac{\cos(\varphi_l - \varphi_m - \varphi_s + \varphi_{m+s-l})}{(l-m)^2 \Omega^2 + \left(\gamma + \Gamma^{-1} \sum_j F_j^2\right)^2} \right\}, \tag{33}$$

from which it is apparent that the absorption depends on the phases φ_m.

To exhibit the phase relations for which the losses of absorbed energy are minimized we integrate expression (33) over the filter length. Assuming that the filter length is much greater than the emission wavelength but much smaller than the cavity length L, and neglecting the phase-independent part of the absorption energy, the problem reduces to an analysis of the minimum of the expression

$$\Phi(\varphi) \equiv -\sum_s \sum_l \sum_m A_m A_l A_s A_{s+l-m} (l-m)^2 \left[1 + \cos 2\pi \frac{X}{L}(m-l) + \cos 2\pi \frac{X}{L}(m-s)\right] \cos(\varphi_m - \varphi_l - \varphi_s + \varphi_{m+s-l}), \tag{34}$$

in which X is the distance from the center of the filter to the cavity mirror.

In the event the filter is located at the edge of the cavity, X/L = 0, the quantity $\Phi(\varphi)$ is minimized when the phases satisfy the relation

$$\varphi_{m+1} - \varphi_m = \varphi_m - \varphi_{m-1} + 2\pi v, \tag{35}$$

in which v is an integer.

If the filter is in the center of the cavity, X/L = 1/2, then $\Phi(\varphi)$ is minimized by the phase relations

$$\varphi_{m+1} - \varphi_m = \varphi_m - \varphi_{m-1} + \pi(2v+1). \tag{36}$$

The analysis of the minimization of $\Phi(\varphi)$ for X/L = 1/3 leads to the following phase relations:

$$\varphi_{m+3} - \varphi_m = \varphi_{l+3} - \varphi_l = 2\pi v. \tag{37}$$

It must be emphasized that relations (35)-(37) minimize the losses for any amplitudes A_m. But if we consider arbitrary values of the filter coordinates the phase relations will depend on A_m and be extremely complicated.

For cases in which condition (35) or (36) is satisfied the phases of all components can be expressed in terms of two arbitrary constants (φ_0 and α):

$$\varphi_m = \varphi_0 + m\alpha \qquad (X/L = 0), \tag{38}$$

$$\varphi_m = \varphi_0 + m\alpha + \frac{1}{2} m(m-1)\pi \qquad \left(X/L = \frac{1}{2}\right). \tag{39}$$

In order to write the phases of all the components in the case of (37) it is necessary to introduce four constants (φ_0, α, β, θ):

$$\varphi_{3m} = \varphi_0 + m\theta, \quad \varphi_{3m+1} = \varphi_0 + \alpha + m\theta, \quad \varphi_{3m+2} = \varphi_0 + \beta + m\theta \quad (X/L = {}^1/_3). \tag{40}$$

The time dependence of the emitted radiation will, in general, differ in these three cases. The low-frequency component of the field intensity oscillations, which is registered by a square-law detector, is proportional to the quantity

$$B = \sum_l \sum_m A_m A_l \cos[(m-l)\Omega t + \varphi_m - \varphi_l]. \tag{41}$$

If the phases of one of the relations (38)-(40) is specified, the above quantity is written as follows:

$$B = \sum_m \sum_l A_l A_{l+m} \cos[m(\Omega t + \alpha)] \quad \left(\frac{X}{L} = 0\right), \tag{42}$$

$$B = \sum_m \left\{ \sum_l (-1)^l A_l A_{l+2m+1} \cos[(2m+1)(\Omega t + \alpha)] - \sum_l A_l A_{l+2m} \cos[2m(\Omega t + \alpha)] \right\} \quad \left(\frac{X}{L} = \frac{1}{2}\right), \tag{43}$$

$$B = \sum_m \Big\{ \sum_l A_l A_{l+3m} \cos[3m(\Omega t + \alpha)] + \sum_l A_{3l} A_{3(l+m)+1} \cos[(3m+1)\Omega t + \alpha + m\theta] +$$

$$+ \sum_l A_{3l+1} A_{3(l+m)+2} \cos[(3m+1)\Omega t + \beta - \alpha + m\theta] + \sum_l A_{3l+2} A_{3(l+m+1)} \cos[(3m+1)\Omega t + (m+1)\theta - \beta] +$$

$$+\sum_l A_3\, A_{3\,(l+m)+2} \cos\left[(3m+2)\,\Omega t+\beta+m\theta\right]+\sum_l A_{3l+1} A_{3\,(l+m+1)} \cos\left[(3m+2)\,\Omega t+(m+1)\,\theta-\alpha\right]+$$

$$+\sum_l A_{3l+2} A_{3\,(l+m)+4} \cos\left[(3m+2)\,\Omega t+(m+1)\,\theta+\alpha-\beta\right]\Big\}\quad \left(\frac{X}{L}=\frac{1}{3}\right). \tag{44}$$

In the case X/L = 0, therefore, all terms having the same frequency have identical phases. In the case X/L = 1/2 all even-frequency terms have the same phase, while some of the odd-frequency terms are in phase opposition with the others. In both of these cases, as apparent from (42) and (43), if the amplitudes A_m are known, the form of the time dependence of the signal is completely determined. The indeterminate constant α merely gives a slight phase shift.

In the third case the form of the output signal depends on the constants α, β, and θ, which are not defined in the given model. The reason for the latter situation is that with the cell in this position (X/L = 1/3) the losses are independent of α, β, and θ, and all of these constants remain arbitrary. It is possible that in an actual system a definite relationship is established between α, β, and θ by the material of the active rod, whose role was not taken into account in the given case. It is seen from Eq. (44) that for any values of the constants α, β, and θ the output signal will have intense components at the frequency 3Ω and at integer multiples thereof.

§2. Dependence of the Self-Locking Effect on the Position of the Bleachable Filter in the Cavity

In our investigation of the self-locking effect we set as our objective to explain the fundamental laws of the effect and their dependence on the laser parameters, to verify the theoretical results, and on the basis of a comparison of the experimental data with the theory to try and portray the general pattern of the physical processes involved in self-locking.

We mainly studied the dependence of the time behavior of the laser emission with self-locking on the position of the bleachable filter inside the cavity. A diagram of the experimental arrangement is shown in Fig. 35. Here R_1 and R_2 are dielectric cavity mirrors on plane-parallel substrates with reflectivities of 99 and 65%, respectively, at the wavelength λ = 1.06 μ; NG is a KGSS-7 neodymium glass rod 15 mm in diameter and 120 mm long; K is a cell containing a solution of No. 7 or No. 4 dye in nitrobenzene, forming the bleachable filter; SM is a semitransparent mirror; PhEC is a coaxial photoelectric cell (type FÉK-09) connected directly to the deflection plates of an S1-11 oscilloscope (time constant for the total system: ~2 nsec); FP is a Fabry–Perot etalon with plate separation d_{FP} = 15 cm; L is a lens with focal length f = 1000 mm; P is a photosensitive plate; and D is a photodiode connected to the input of an S1-16 oscilloscope (time constant of this system: 1 to 3 μsec). The system afforded the possibility of simultaneously recording the temporal and spectral characteristics of the laser output radiation. The photodiode D was used to control the number of giant pulses per single pumping pulse. Also, the emission spectrum was investigated with a diffraction spectrograph having a linear dispersion dλ/d ≈ 1.25 Å/mm. The optical length of the cavity was L = 300 to 320 cm, and the initial transmissivity of the bleachable filter was made equal to T_0 = 0.50 to 0.85.

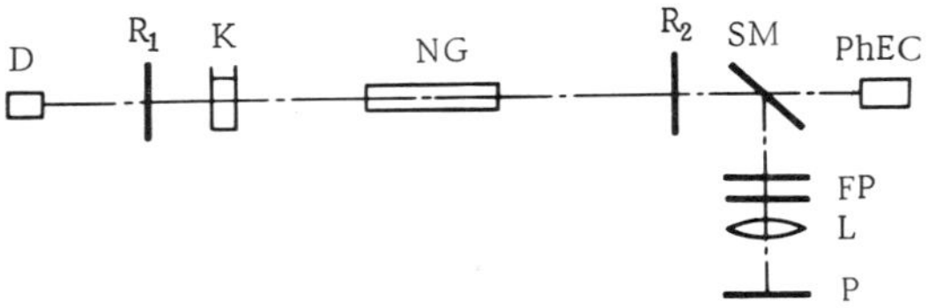

Fig. 35. Diagram of experimental arrangement.

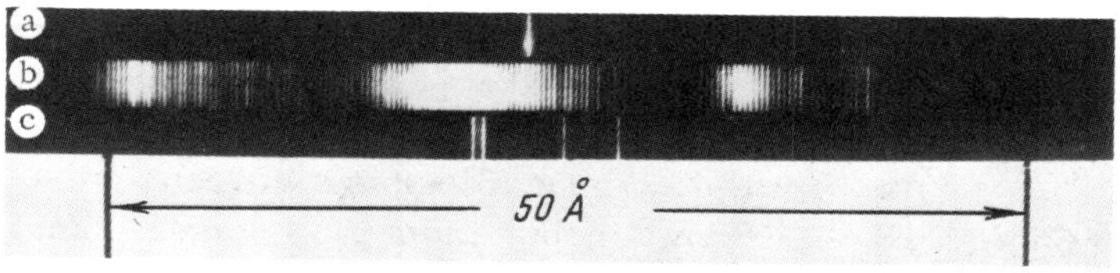

Fig. 36. Spectrograms of emission from a neodymium glass laser (L = 324 cm). a) Single giant pulse with passive switching, T_0 = 65%; b) free oscillation (at five times the pumping threshold); c) four giant pulses with passive switching, T_0 = 65%.

In order to facilitate the calculations according to Eqs. (42)-(44) and the comparison of the theoretical results with the experimental data we chose the laser parameters so that only a few modes would be excited. For this purpose the mirrors R_1 and R_2 forming the cavity were on plane-parallel substrates, and the cell K containing the bleachable solution was oriented at an angle $\leqslant 1°$ relative to the cavity axis. Moreover, an active rod with its end faces perpendicular to the cavity axis was used. As a result of the combined selective action of all the cavity elements the emission spectrum for near-threshold pumping signals (i.e., with the emission of a single giant pulse) comprised a single line whose width is determined by the instrument function of the equipment (Fig. 36a). With an increase in the pumping energy the number of giant pulses (during a single pumping signal) increases, but the time variation of the first pulse does not change. The emission spectrum in this case (Fig. 36c) generally consists of several intense lines, the number of which corresponds to the number of giant pulses and the width of which is determined by the instrument function (~ 0.1 Å). It is important to note that the emission spectrum of both the single giant pulse and several pulses can contain other lines with intensities one to three orders of magnitude less than that of the principal lines, so that they can only be observed on the spectrogram if the latter is greatly overexposed. Measurements of the width of the spectrum for a single giant pulse by means of the Fabry–Perot etalon (instrument function: $\delta\nu_{FP} \leqslant 0.3 \cdot 10^{-2}$ cm^{-1}) showed that this quantity varies over the range $2\Delta\nu_{osc}$ = $(0.7$ to $1.2) \cdot 10^{-2}$ cm^{-1}, the variations of the linewidth in this range bearing a random character unrelated to any variation of the position of the cell K in the cavity.*

To eliminate the discriminating action of the ends of the neodymium glass rod on the Q-values of closely spaced modes the axis of the rod was tilted 0.5 to 1.0° from the cavity axis in all the experiments. Otherwise the spectrum of the excited modes and, hence, the time variation of the emitted radiation would depend on the position of the active rod, a situation that could render difficult the interpretation of the results pertaining to the dependence of the time pattern on the position of the bleachable filter.

The principal experimental results may be summarized as follows.

For large values of the filter transmissivity (T_0 = 85 to 90%) and with the filter cell situated near one of the mirrors the giant pulse is intensity-modulated at a frequency $f_1 = c/2L$ (Figs. 37a and 37b), with the percentage modulation not exceeding 80% and the general behavior changing from one burst to the next but remaining nearly sinusoidal. With the cell positioned in the center of the cavity the percentage modulation is decreased, and the frequency is equal in some cases to $f_2 = 2(c/2L)$ and in others to $f_1 = c/2L$ (Figs. 37c and 37d). The emission

* The Fabry–Perot etalon measurements were conducted only in the case of a single giant pulse.

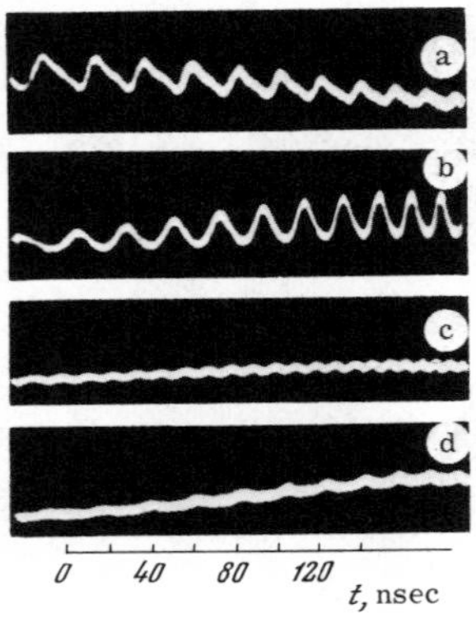

Fig. 37. Oscillograms of giant-pulse emission from a neodymium glass laser (sample dimensions 15 mm diameter × 120 mm) with a passive shutter (No. 4 dye in nitrobenzene). $L = 324$ cm; $L_1 = 20$ cm; rod tilted ~7' relative to cavity axis; $T_0 = 88\%$. a, b) $X_2 = 5$ cm; c, d) $X_2 = 162$ cm.

spectrum was not investigated in these cases, due to the extreme difficulty of obtaining a single giant pulse for such large values of T_0.

A reduction of the filter transmissivity to $T_0 \sim 75\%$ radically alters the percentage modulation, which is now close to 100%. Typical time scans obtained for a pulse with different positions of the filter in the cavity are shown in Fig. 38. For X_1 (or X_2) ≈ 2 cm the giant pulse has the form of a train of separate narrow pulses. Moving the filter to the position $X_1 = X_2 =$ 162 cm doubles the subpulse repetition rate; the transition from the repetition frequency $f_1 = c/2L$ to the double frequency $f_2 = 2(c/2L)$ during the pulse buildup time is evident in Fig. 38c. In the case of X_1 (or X_2) = 108 cm the modulation has a more complex form (particularly in Figs. 38e and 38g), but it exhibits a dominant frequency $f_3 = 3(c/2L)$.

The most stable (in the sense of the character of the modulation in transition from one burst to another) time scans, with a percentage modulation close to 100%, are obtained with a further reduction in the transmissivity of the filter. The case for $T_0 = 65\%$ is illustrated in Fig. 39. The total width of the giant pulse (envelope) in this case is 100 to 150 nsec, and the output energy is ~0.3 J.* For the case of X_1 (or X_2) = 2 cm (Fig. 39a), when the width of the individual subpulses is equal to T_{pls} = 3 or 4 nsec and the repetition period $T_{rep} = 22 \pm 1$ nsec ≈ $2L/c$, the peak power attains values P_{max} = 15 to 20 MW.

With the cell in the center of the cavity the distance between two adjacent pulses is $T_{rep} = 11 \pm 1$ nsec (Fig. 39b). The width of the individual pulses in this case is again equal to T_{pls} = 3 or 4 nsec, and the relative intensity of two adjacent pulses is usually not equal to unity, but is almost constant for the entire train. The peak power decreases in correspondence with the reduction in the number of pulses to P_{max} = 7 to 10 MW.

When the filter cell is located at a distance X_1 (or X_2) = 108 cm ≈ $L/3$, the emitted radiation (Fig. 39c) consists of a train of pulses with separation $T_{rep} = 7 \pm 1$ nsec ≈ $2L/3c$. As in the preceding case, the relative intensity of three adjacent pulses can vary from one burst to the next, but remains constant over the entire train. The absolute value of the peak power in this case decreased to 5 to 7 MW.

Consequently, the first position of the cell, near one of the cavity mirrors, such that the pulse spacing in the train and their peak power are maximal, is the most significant from the standpoint of practical applications.

To compare the resulting experimental dependences of the time variation of the giant pulse on the filter position in the cavity with the theory according to Eqs. (42)-(44) we plotted

*It is essential to note that for such large cavity lengths ($L \approx 300$ cm) emission occurs only in a small portion of the active rod cross section (~0.15 cm^2).

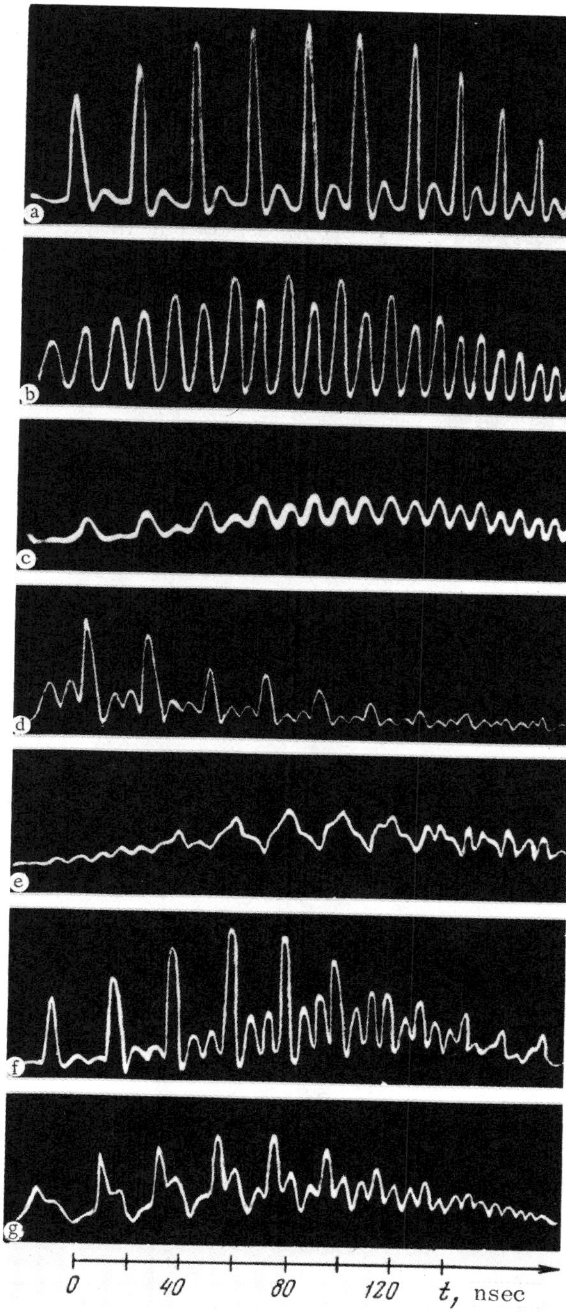

Fig. 38. Oscillograms of giant-pulse emission from a neodymium glass laser (sample dimensions 15 mm diameter × 120 mm) with a passive shutter (No. 4 dye in nitrobenzene). L = 134 cm; L_1 = 20 cm; rod tilted ~ 7' relative to cavity axis; T_0 = 75%. a) X_1 = 2 cm; b, c) $X_1 = X_2$ = 162 cm; d,e) X_2 = 108 cm; f, g) X_1 = 108 cm.

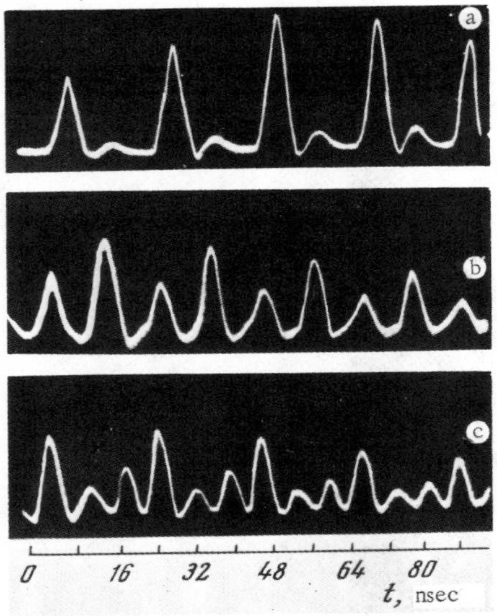

Fig. 39. Oscillograms of giant-pulse emission from a neodymium glass laser (sample dimensions 15 mm diameter × 120 mm) with a passive shutter (No. 7 dye in nitrobenzene). L = 324 cm; L_1 = 20 cm; rod tilted ~10' relative to cavity axis; T_0 = 65%. a) X_2 = 2 cm; b) $X_1 = X_2$ = 162 cm; c) X_2 = 216 cm.

graphs of the output signal intensity versus time for the three filter positions. It was assumed on the basis of an experimental determination of the emission linewidth that only seven of the spectral components have nonzero amplitudes: $A_m \neq 0$ for $|m| \leq 3$, where all of these amplitudes were regarded as equal. In the case X/L = 1/3 the following additional phase relations were also specified: $\beta = 2\alpha + \pi$; $\theta = 3\alpha$.

The calculated time dependence of the signal intensity is shown in Fig. 40. With the filter near the edge of the cavity the distance between intensity maxima on the graph is $2\pi/\Omega = 2L/c$; it is divided in half with the filter placed at the center, and into one third for the filter coordinate X = L/3. The same pulse separations were also obtained experimentally (cf. Fig. 39).

Consequently, the experimental results exhibit good agreement with the theoretical results for the time dependence of the emitted radiation intensity for different positions of the bleachable filter. This, in turn, indicates that the postulated theoretical model adequately describes the self-locking mechanism.

However, the intensity ratios of adjacent pulses for X = L/2 and X = L/3 differ somewhat between the theoretical curves and the experimental oscillograms; this disparity is attributable, clearly, to certain simplifications allowed in the model. The asymmetry of the cavity should

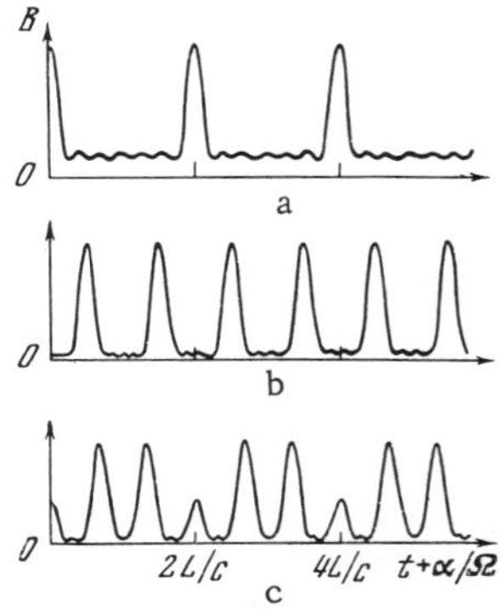

Fig. 40. Output signal intensity B (in relative units) versus time according to Eqs. (42)-(44). a) X/L = 0; b) X/L = 1/2; c) X/L = 1/3.

probably be taken into account for better matching of the theoretical and experimental curves; this is particularly true of the difference between the losses at the mirrors R_1 and R_2 [allowance for the mirror losses changes the field expression (31)].

As we noted earlier in this section, a theoretical and experimental investigation of the same problem [92] was published concurrently with our own work, but in application to a ruby laser. The results (theoretical and experimental) obtained in the cited paper are consistent with our own, but the authors of [92] confined both their theory and experiment to two filter positions ($X/L = 0$ and $1/2$) and thus did not obtain the behavior patterns observed for $X/L = 1/3$. The investigation of the latter case, however, is very significant, because the nature of the modulation for $X/L = 1/3$ fosters the conclusion that placement of the filter in positions $X = L/m$ increases the pulse repetition frequency $f_{rep} = mf_1$, rather than doubling the number of pulses with a separation between the latter that is a multiple of the distance to the nearest mirror. Similar results were obtained later in [93, 94], in which the authors observed pulse repetition frequencies in ruby and neodymium glass lasers with a bleachable filter from $f_1 = c/2L$ to $f_8 = 8(c/2L)$ when the filter was mounted in positions $X = L/m$.

§3. Self-Locking in Neodymium Glass and Ruby Lasers with a Bleachable Filter in the Case of a Broad Emission Spectrum; Ultimately Narrow Width of Ultrashort Pulses in Lasers of This Type

As shown in the preceding section, with the bleachable filter set up in the position $X/L = 0$ the simplest type of phase self-locking is realized. This type of mode locking is the best suited to practical application of the given effect for the generation of ultrashort pulses having a high peak power and large spacing between pulses (off-duty factor). This is easily demonstrated in the elementary example of the emission spectrum of N_m locked modes of equal amplitude. According to (42), the field intensity registered by a square-law detector in the case $X/L = 0$ is proportional to the quantity

$$B=\sum_m\sum_l A_l A_{l+m}\cos\left[m(\Omega t+\alpha)\right]=\Big(\sum_m A_m\cos m\Omega t\Big)^2. \tag{45}$$

Setting $A_m = A_0$ for $|m| \leq N_m/2$, we obtain the function

$$B=A_0\left[\frac{\sin\left(\frac{N_m}{2}\Omega t\right)}{\sin\left(\frac{1}{2}\Omega t\right)}\right]^2, \tag{46}$$

which represents a periodic (period $T_{rep} = 2L/c$) train of individual narrow pulses with a peak power proportional to $B_{max} = N_m^2 A_0^2 = N_m^2 B_1 = N_m B_{av}$. Here B_1 is the power of one mode, and $B_{av} = N_m A_0^2$ is the average power for the N_m unsynchronized modes. Consequently, the peak power with the synchronization (locking) of N_m modes is increased N_m-fold. As for the individual pulse width T_{pls}, for $N_m \gg 1$ it is approximately equal to

$$T_{pls}\approx\frac{2.78\cdot 2L}{\pi c N_m}=\frac{0.88}{c\cdot 2\Delta\nu_{em}}\approx\frac{T_{rep}}{N_m}, \tag{47}$$

where $2\Delta\nu_{osc}$ is the total width of the spectrum of locked modes (in cm^{-1}).

As evident from Eqs. (46) and (47), the self-locking method affords the greatest advantages for increasing the peak power and obtaining ultrashort pulses in lasers that have a fairly broad emission spectrum. Crude estimates of the capabilities of the given method for the best-known ruby lasers ($2\Delta\nu_{\text{lum}} \approx 10$ cm^{-1} [95]) and neodymium glass lasers ($2\Delta\nu_{\text{lum}} \approx 250$ cm^{-1} [66, 87]) show that in the case of the ruby laser it is reasonable to expect pulse widths on the order $T_{\text{pls}} \sim 3 \cdot 10^{-12}$ sec, and in the case of neodymium glass $T_{\text{pls}} \sim 10^{-13}$ sec. It is not surprising that such impressive potential capabilities as offered by the self-locking effect have stimulated numerous researchers in recent years to try and achieve these results.

The main problems in want of solution in this connection were the following: 1) to obtain the broadest possible laser emission spectrum in the self-locking regime, since, other requirements having been met, it is the spectrum width that limits the width of the emitted ultrashort pulses; 2) to eliminate any possible mode discrimination in the laser cavity at various elements, since this type of discrimination, besides possibly narrowing the spectrum [22], tends to limit the off-duty factor of the emitted pulses [96, 97]; 3) to have the laser with a bleachable filter operate in an emission regime such that all excited modes are phase-synchronized, i.e., locked; 4) to measure the width of the emitted pulses.

We investigated the first two problems (see [22, 23, 82]), making it possible to a certain extent to shed light on the stated aspect of the effect and to formulate recommendations for the use of a particular laser cavity design in connection with a bleachable filter. We must also bring attention to [97], in which the authors infer on the basis of a comparison of the results from an investigation of the temporal and spectral characteristics of ultrashort pulses that mode discrimination at various cavity elements plays an important role in experiments of this nature. The third problem was treated at about the same time in the theoretical papers [98-100], which allowed certain conclusions to be drawn with regard to the requirements on the emission regime [98, 100] and on the properties of the active medium and bleachable filter [99, 100] from the point of view of generating pulses of minimum width. Some of these results have been corroborated experimentally to some extent [101]. Also to be included in this group of papers are the experimental studies [97, 102, 103], which were concerned with realization of the self-locking regime in ruby and neodymium glass lasers with a bleachable filter, for which, evidently, all excited modes are locked. Nothing is said, however, about the characteristics of the regime itself in this case. Inasmuch as these latter papers were based on measurements of the widths of the emitted pulses, they automatically solved the fourth problem stated above. We should nevertheless cite in addition the earliest work on this problem [104], as well as [105, 106].

In the present section we consider cavity designs for neodymium glass and ruby lasers with a bleachable filter, for which continuous emission spectra of maximum width are obtained and stable temporal characteristics are observed on the part of the giant pulse, evincing self mode locking. Moreover, we also examine the problem of the giant-pulse buildup process in a laser using a bleachable filter and its effect on the total width of the emission spectrum.

The experiment was carried out with neodymium glass and ruby lasers [23]. In the neodymium glass case the passive shutter was a solution of No. 7 dye in nitrobenzene, and in the ruby case a solution of cryptocyanine in ethanol.* The cavity structures of both lasers were such that mode discrimination at internal elements was eliminated. As our investigations showed, the most stable (from the point of view of reproducibility) temporal patterns of the laser emission are observed with the separation of a single low-order angular mode (to modes TEM_{00p}). For this we placed two diaphragms (2 mm in diameter for the neodymium glass laser and 4 mm for the ruby) near the ends of the active rod, where they increased the diffraction losses for higher-order angular modes.

---

* Some of the neodymium glass laser experiments were also conducted using No. 1 and No. 4 dyes, but no appreciable differences were noticed in the final results. In the case of the ruby laser a solution of cryptocyanine in nitrobenzene was also used as the passive shutter, and similar results were obtained.

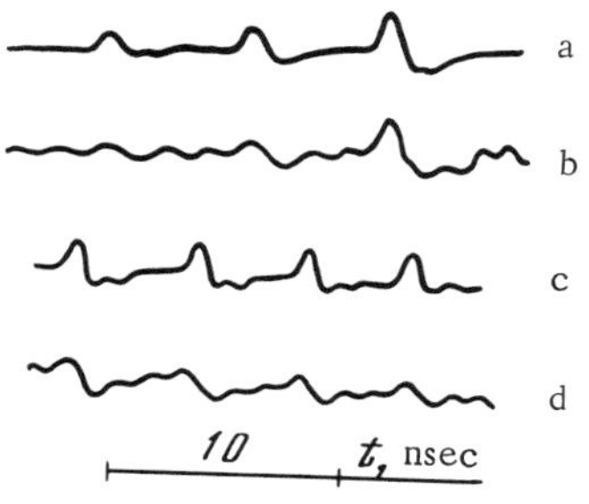

Fig. 41. Oscillograms of giant-pulse emission from a laser with a passive shutter. a, b) Neodymium glass laser (sample dimensions 16 mm diameter × 220 mm); passive shutter: No. 7 dye in nitrobenzene, $T_0 = 65\%$, $L = 99$ cm; c, d) ruby laser (sample dimensions 15 mm diameter × 120 mm), passive shutter: cryptocyanine in ethanol, $T_0 = 45\%$, $L = 85$ cm; a, c) new mirrors; b, d) after several (three or four) bursts.

With this cavity design and the cell containing the bleachable solution positioned near the mirror R_1 ($\alpha \approx \alpha_B$) the temporal characteristics of the emission from the neodymium glass and ruby lasers have the form illustrated in Figs. 41a and 41c. The giant pulse in this case represents a train of equidistant narrow pulses, whose width $T_{pls} = 0.8$ to 1.2 nsec is determined by the resolving power of the detection system: an FÉK-09 photocell and I2-7 oscilloscope. The pulse separation, within the experimental error limits, is equal to $T_{rep} = 2L/c$, and the total width of the train (at the 0.5-peak level) usually comes to 70 to 150 nsec. The reproducibility of these patterns for the given experimental arrangement is no worse than 80 to 90%. But it is essential to bear in mind here that after every two to four bursts the cavity mirrors (usually the exit mirror R_2) begin to suffer from hole-burning, which deteriorates both the emission patterns themselves (Figs. 41b and 41d) and their stability. This deterioration of the temporal characteristics is possibly associated with a change in the orders of the excited angular modes in the course of emission due to hole-burning in certain regions of the cavity mirrors. This change in the angular orders can eventually upset (in the course of emission) the mode-locking conditions.

As for the giant-pulse emission spectra in the self-locking state, which we investigated simultaneously with the temporal characteristics [106], they are continuous, correct to the resolution of the spectral instruments used (~ 0.1 cm^{-1}). In order to estimate the ultimate width T_{lim} that can be obtained for ultrashort pulses with self-locking of all modes of the spectrum, we studied the intensity distribution in the emission spectrum of the neodymium glass and ruby lasers, because with the locking of all excited modes it is the width and shape of this distribution that governs the ultimate width of the ultrashort pulses. The emission spectrum of the neodymium glass laser was recorded with a diffraction spectrograph having a dispersion $d\lambda/d \approx 2.5$ Å/mm. The ruby laser spectrum was investigated by means of a Fabry–Perot interferometer having a plate separation $t_{FP} = 2$ mm. The resolution in both cases was ~ 0.1 cm^{-1}. Along with the giant-pulse emission spectra, optical density reference marks were imposed on the same film by photographing a series of emission spectra in the free-oscillation regime, for which a constant output energy is ensured, within 10% limits, for the laser emission from one burst to next. The spectral intensity was varied by means of calibrated neutral light filters. Inasmuch as the total width of the giant pulse in our experiments was $\sim 10^{-7}$ sec and the width of the free-oscillation spikes was $\sim 10^{-6}$ sec, we were able to process the investigated spectrum by means of the resulting optical density curves, using ordinary photographic photometry techniques.

Typical intensity distributions in the emission spectra of a neodymium glass and a ruby laser with a passive shutter in the self-locking regime are shown in Figs. 42a and 42b, respectively. As the figures reveal, the distribution for both lasers has a nearly Gaussian form. The width $2\Delta\nu_{osc}$ of the spectrum (at the 0.5-peak level) changes slightly from one pulse to another, falling within the limits $2\Delta\nu_{osc} = 15$ to 30 cm^{-1} for the neodymium glass laser and $2\Delta\nu_{osc} = 0.6$ to 1.3 cm^{-1} for the ruby laser.

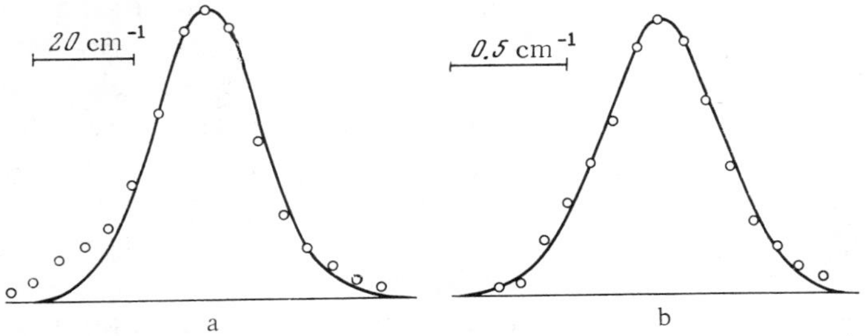

Fig. 42. Typical examples of spectral distributions of the emission intensity (in relative units) for lasers with a passive shutter in the self-locking regime. a) Neodymium glass laser; b) ruby laser. The curves represent Gaussian distributions for the corresponding values of $2\Delta\nu_{osc}$, and the dots represent the experimental data.

The width of the giant-pulse emission spectrum for a laser with a passive shutter can also be roughly estimated on the basis of the notions set forth in Chapter III. Accordingly, the giant-pulse buildup process in this type of laser can be conditionally divided into two intervals. The first interval (from the inception of emission to the initiation of bleaching of the filter) is characterized by a relatively slow growth of the photon density in the cavity due to the slight excess of the gain of the active medium over the total losses. Mode coupling need not be taken into account on this interval, since the shutter operates as a linear element at this time. The second interval (from the beginning of bleaching to the time at which the pulse power is a maximum) is characterized by a far more rapid (relative to the first interval) growth of the photon density and mode interaction due to the nonlinear character of the absorption in the shutter, resulting in mode self-locking. The length of the first interval is usually an order of magnitude greater than the length of the second interval, so that the first interval may be regarded as responsible for the total giant-pulse buildup time.

At present it appears impossible to take account of the possible influence of mode coupling during the bleaching period on the resultant emission spectrum, on account of the very large number of modes excited and the rather complex character of this coupling. We shall therefore estimate the influence of only the first interval of the giant-pulse buildup period on the total width of the emission spectrum and from a comparison of this estimate with the experimental value obtain information concerning the influence of the second interval on the resultant spectrum.

To take account of the influence of the first interval, following [83], we express the ratio of the gains (during the round-trip transit time t_c of a photon across the cavity) at the maximum ν_a of the luminescence line and at a frequency ν in the form

$$\frac{G(\nu_a)}{G(\nu)} = \frac{R_{1a} R_{2a} T_a^2 \exp(2\sigma_a N l_a)}{R_{1\nu} R_{2\nu} T_\nu^2 \exp[2\sigma_a(\nu) N l_a]}, \tag{48}$$

in which the notation is the same as in Chapter III. Assuming that the amplification (or luminescence) line has a dispersion form and width $2\Delta\nu_a$ (at the 0.5-peak level), $\sigma_a(\nu) = \frac{\sigma_a}{1 + [(\nu - \nu_a)\,\Delta\nu_a]^2}$, and making use of the fact that the width of the emission spectrum for ruby and neodymium glass lasers $2\Delta\nu_{osc} \ll 2\Delta\nu_a$ and is much less than the absorption bandwidth of the passive shutter, we can assume approximately that $R_{1\nu} \approx R_{1a}$, $R_{2\nu} \approx R_{2a}$, $T_\nu \approx T_a$, and

$$\frac{G(\nu_a)}{G(\nu)} \approx 1 + 2\sigma_a N l_a \frac{[(\nu - \nu_a)/\Delta\nu_a]^2}{1 + [(\nu - \nu_a)/\Delta\nu_a]^2}. \tag{49}$$

But the intensity ratio between the maximum of the luminescence line and the frequency ν at the inception of bleaching is equal to*

$$\frac{P(\nu_a)}{P(\nu)} = \left[\frac{G(\nu_a)}{G(\nu)}\right]^q \approx \left\{1 + 2\sigma_a N l_a \frac{[(\nu - \nu_a)/\Delta\nu_a]^2}{1 + [(\nu - \nu_a)/\Delta\nu_a]^2}\right\}^q, \tag{50}$$

where q is the time from the initiation of emission to the instant of bleaching, expressed in terms of the number of round-trip photon transits across the cavity. The quantity $2\sigma_a N l_a$ in this expression can be determined from the stimulated emission threshold condition $R_{1a} R_{2a} T_a^2 \cdot \exp[2\sigma_a N l_a] = 1$. Setting $P(\nu_a)/[P(\nu_a \pm \delta\nu_{1/2}] = 2$, we can determine the frequency $\nu_a \pm \delta\nu_{1/2}$ at which in the emission spectrum the radiation intensity must be half the value at the maximum. By the definition of the emission line width, it must be equal to $2\Delta\nu_{osc} = 2\delta\nu_{1/2}$.

In our experiments with a neodymium glass laser $R_1 \approx 1.0$, $R_2 = 0.55$, $T_a = 0.6$, $2\Delta\nu_a = 250\ cm^{-1}$, $2\sigma_a N l_a \approx 1.4$, and according to our measurements in [89] $q \lesssim 700$. The insertion of these values into (50) yields $\Delta\nu_{osc}/\Delta\nu_a \approx 1/40$, or $2\Delta\nu_{osc} \approx 6\ cm^{-1}$. For the ruby laser, accordingly, we have $R_1 = 0.89$, $R_2 = 0.55$, $T_a = 0.40$, $2\Delta\nu_a = 10\ cm^{-1}$, $2\sigma_a N l_a \approx 2.5$, and according to the data of [88] $q \approx 1150$, which gives $\Delta\nu_{osc}/\Delta\nu_a \approx 1/70$, or $2\Delta\nu_{osc} \approx 0.15\ cm^{-1}$.

It is evident from Fig. 42 that the measured widths of the giant-pulse emission spectra for lasers with a passive shutter, as expected, are considerably smaller (by an order of magnitude) than the widths of the luminescence lines of the corresponding active media. However, the experimental values are still several times larger than the values obtained by the foregoing calculation, which only takes account of the initial stage (up to the beginning of bleaching) of the giant-pulse buildup process. We attribute this discrepancy to the influence of mode interaction during the bleaching of the shutter, so that in a certain sense it is a quantitative estimate of the influence of this interaction on the resultant emission spectrum. Support for this hypothesis may be found in the fact that the width of the emission spectrum in the single-spike free-oscillation regime (when the mode coupling associated with nonlinear absorption in the passive shutter is absent) is less than the width of the giant-pulse spectrum (see Fig. 25) and is close to the calculated value. On the other hand, the transient process for the free-oscillation spike should clearly not differ significantly from the initial stage (up to the initiation of bleaching) of the giant-pulse transient in a laser with a passive shutter under identical pumping conditions (cf. the results of the measurements of q in §§2 and 3 of Chapter III for a giant pulse and a single free-oscillation spike).

We now estimate the ultimate width of ultrashort pulses on the assumption that the intensity distribution in the giant-pulse emission spectra is described by a Gaussian curve and that all modes of the spectrum are locked-in. The low-frequency component of the field intensity in this case is expressed as follows:

$$B = \Big(\sum_{m=-N_0/2}^{N_0/2} A_m \cos m\Omega t\Big)^2 = \Big\{\sum_m A_0 \exp\Big[-\Big(\frac{2m}{N_m}\Big)^2 \frac{\ln 2}{2}\Big] \cos m\Omega t\Big\}^2, \tag{51}$$

---

* Allowance is not made here for the fact that, due to the frequency dependence $\sigma_a(\nu)$, the oscillation initiation times differ for different modes [83]. Allowance for this fact, however, is of little consequence for our estimate.

where N_0 is the total number of modes in the spectrum and N_m is the number of modes contained within the width $2\Delta\nu_{em}$ of the emission spectrum. The expression obtained for B as a function of the time represents a train of equidistant pulses with period $T_{rep} = 2\pi/\Omega = 2L/c$. Considering that for the estimation of the pulse width it suffices to limit the summation to the small region of variation of the parameter $|\Omega t| < \pi/N_m$, we can replace the summation by integration. But since in our case $N_0 > 10N_m$, the presence of the exponential factor in (51) permits us to extend the limits of integration to $\pm\infty$, so that (as shown by estimate) only an insignificant correction is obtained.

We obtain as a result of integration

$$B = A_0^2 \left\{ \int_{-\infty}^{\infty} \exp\left[-\left(\frac{2z}{N_m}\right)^2 \frac{\ln 2}{2}\right] \cos z\Omega t\, dz \right\}^2 = A_0^2 \frac{\pi N_m^2}{2 \ln 2} \exp\left[-\frac{\Omega^2 t^2 N_m^2}{4 \ln 2}\right], \tag{52}$$

whence the ultimate pulse width (at the 0.5-peak level) is equal to

$$T_{\lim} \approx \frac{0.44 T_{rep}}{N_m} = \frac{0.44}{c \cdot 2\Delta\nu_{em}}. \tag{53}$$

Substituting our measured values for $2\Delta\nu_{em}$, we obtain for the ultimate width of ultrashort pulses emitted by a neodymium glass laser $T_{\lim}(Nd^{3+}) = (0.5 \text{ to } 1.0) \cdot 10^{-12}$ sec, and for a ruby laser $T_{\lim}(Cr^{3+}) = (1.1 \text{ to } 2.4) \cdot 10^{-11}$ sec.

It is important to note that our estimates of the lower-limiting width are close to the experimental data obtained by several authors (under roughly the same experimental conditions: for a neodymium glass laser $T_{exp}(Nd^{3+}) = (0.7 \text{ to } 2.0) \cdot 10^{-12}$ sec [97, 103, 105], and for a ruby laser $T_{exp}(Cr^{3+}) = (1.2 \text{ to } 1.4) \cdot 10^{-11}$ sec [102, 103].

I am pleased, in conclusion, to express gratitude to my scientific director, Prof. V. I. Malyshev, for his constant assistance and interest in the present study. I also wish to thank Prof. M. M. Sushchinskii for reviewing the manuscript and offering valuable comments. And I should like to thank V. S. Petrov, T. I. Kuznetsova, A. A. Sychev, and A. V. Masalov for assisting with and discussing certain investigations, as well as Yu. S. Ivanov for helping with several of the experiments.

LITERATURE CITED

1. N. G. Basov, V. S. Zuev, and P. G. Kryukov, Zh. Éksp. Teor. Fiz., 43:353 (1962).
2. B. P. Stoicheff, Internat. School of Physics "Enrico Fermi," Course XXXI, August 19-31 (1963).
3. F. J. McClung and R. W. Hellwarth, J. Appl. Phys., 33:828 (1962).
4. F. S. Barnes, Proc. IRE, 50:1686 (1962).
5. J. L. Helfrich, J. Appl. Phys., Part I, 34:1000 (1963).
6. A. J. De Maria, J. Appl. Phys., 34:2984 (1963).
7. A. J. De Maria, R. Gagosz, and G. Barnard, J. Appl. Phys., 34:453 (1963).
8. A. A. Vuylsteke, J. Appl. Phys., 34:1615 (1963).
9. W. R. Hook, R. H. Dishington, and R. P. Hilberg, Appl. Phys. Lett., 9:125 (1966).
10. J. I. Masters, J. Ward, and E. Hartoni, Rev. Sci. Instr., 34:365 (1963).
11. H. C. Nedderman and Y. Kiang, Proc. IRE, 50:1687 (1962).
12. M. Di Domenico, Jr., J. Appl. Phys., 35:2870 (1964).
13. A. Yariv, J. Appl. Phys., 36:388 (1965).

14. L. E. Hargrove, R. L. Fork, and M. A. Pollack, Appl. Phys. Lett., 5:4 (1964).
15. M. E. Crowell, IEEE J. Quantum Electronics, QE-1:12 (1965).
16. T. Deutsch, Appl. Phys. Lett., 7:80 (1965).
17. A. J. De Maria, C. M. Ferrar, and G. E. Danielson, Jr., Appl. Phys. Lett., 8:22 (1966).
18. V. I. Malyshev and A. S. Markin, Zh. Éksp. Teor. Fiz., 50:339 (1966).
19. H. W. Mocker and R. J. Collins, Appl. Phys. Lett., 7:270 (1965).
20. A. J. De Maria, D. A. Stetser, and H. Heynau, Appl. Phys. Lett., 8:174 (1966).
21. D. A. Stetser and A. J. De Maria, Appl. Phys. Lett., 9:118 (1966).
22. V. I. Malyshev, A. S. Markin, and A. A. Sychev, Zh. Prikl. Spektrosk., 7:662 (1967).
23. V. I. Malyshev, A. S. Markin, and A. A. Sychev, ZhÉTF Pis. Red., 6:503 (1967).
24. P. P. Sorokin, J. J. Luzzi, J. R. Lankard, and G. M. Pettit, IBM J. Res. Devel., 8:182 (1964).
25. P. Kafalas, J. I. Masters, and E. M. E. Murray, J. Appl. Phys., 35:2349 (1964).
26. B. H. Soffer, J. Appl. Phys., 35:2551 (1964).
27. V. N. Gavrilov, Yu. M. Gryaznov, O. L. Lebedev, and A. A. Chastov, Zh. Éksp. Teor. Fiz., 48:772 (1965).
28. F. P. Schäffer and W. Schmidt, Z. Naturforsch., 19a:1019 (1964).
29. B. H. Soffer and R. H. Hoskins, Nature, 204:276 (1964).
30. V. Degiorgio and G. Potenza, Nuovo Cim., B-41:254 (1966).
31. J. A. Armstrong, J. Appl. Phys., 36:471 (1965).
32. A. Szabo and R. A. Stein, J. Appl. Phys., 36:1562 (1965).
33. A. L. Mikaélyan, V. Ya. Anton'yants, V. A. Dolgii, and Yu. G. Turkov, Radiotekh. i Élektron., 10:1350 (1965).
34. O. L. Lebedev, V. N. Gavrilov, Yu. M. Gryaznov, and A. A. Chastov, ZhÉTF Pis. Red., 1(2):14 (1965).
35. V. I. Malyshev, A. S. Markin, and V. S. Petrov, ZhÉTF Pis. Red., Vol. 1, No. 3 (1965).
36. M. P. Vanyukov, O. D. Dmitrievskii, V. I. Isaenko, and V. A. Serebryakov, Dokl. Akad. Nauk SSSR, 167:547 (1966).
37. D. Röss, Z. Naturforsch, 20a:696 (1965).
38. Yu. M. Gryaznov, O. L. Lebedev, and A. A. Chastov, Opt. i Spektrosk., 20:503 (1966).
39. O. L. Lebedev, Yu. M. Gryaznov, A. A. Chastov, and A. V. Kazymov, Zh. Prikl. Spektrosk., 6:261 (1967).
40. G. Bret and F. Gires, Appl. Phys. Lett., 4:175 (1964).
41. V. I. Borodylin, N. A. Ermakova, L. A. Rivlin, and V. S. Shil'dyaev, Zh. Éksp. Teor. Fiz., 48:845 (1965).
42. J. I. Masters, P. Kafalas, and E. M. E. Murray, Bull. Am. Phys. Soc., 9:66 (1964).
43. G. Farkas and I. Kertesz, Phys. Lett., 20:634 (1966).
44. N. T. Melamed, C. Hirayama, and P. W. French, Appl. Phys. Lett., 6:43 (1965).
45. E. Snitzer and R. Woodcock, IEEE J. Quantum Electronics, QE-2:627 (1966).
46. H. W. Gandy, R. J. Ginther, and J. F. Weller, Appl. Phys. Lett., 9:277 (1966).
47. O. R. Wood and S. E. Schwarz, Appl. Phys. Lett., 11:88 (1967).
48. N. V. Karlov, G. P. Kuz'min, Yu. N. Petrov, and A. M. Prokhorov, ZhÉTF Pis. Red., 7:174 (1968).
49. I. V. Grebenshchikov, A. G. Vlasov, B. S. Neporent, and N. V. Suikovskaya, Understanding Optics, Gostekhizdat (1946).
50. M. P. Vanyukov, V. I. Isaenko, and V. A. Serebryakov, Zh. Éksp. Teor. Fiz., 44:1493 (1963).
51. V. I. Malyshev, A. S. Markin, and V. S. Petrov, Zh. Prikl. Spektrosk., 3:415 (1965).
52. S. L. Mandel'shtam, P. P. Pashinin, A. M. Prokhorov, and N. K. Sukhodrev, in: Physics of Quantum Electronics, Conf. Proc. (1966), p. 548.
53. V. V. Korobkin, S. L. Mandel'shtam, P. P. Pashinin, et al., Zh. Éksp. Teor. Fiz., 53:116 (1967).

54. L. D. Derkacheva, A. I. Krymova, V. I. Malyshev, and A. S. Markin, ZhÉTF Pis. Red., 7:468 (1968).
55. V. I. Malyshev, A. S. Markin, V. S. Petrov, I. I. Levkoev, and A. F. Vompe, ZhÉTF Pis. Red., 1(6):11 (1965).
56. W. G. Wagner and B. A. Lenguel, J. Appl. Phys., 34:2040 (1963).
57. A. M. Prokhorov, Radiotekh. i Élektron., 8:1073 (1963).
58. A. L. Mikaélyan and Yu. G. Turkov, Radiotekh. i Élektron., 9:743 (1964).
59. C. C. Wang, Proc. IRE, 51:1767 (1963).
60. V. I. Malyshev and A. S. Markin, Zh. Éksp. Teor. Fiz., 50:1458 (1966).
61. B. A. Ermakov and A. V. Lukin, Zh. Prikl. Spektrosk., 4:410 (1966).
62. B. L. Borovich, V. S. Zuev, and V. A. Shcheglov, Zh. Éksp. Teor. Fiz., 49:1031 (1965).
63. R. McLeary and P. W. Bowe, Appl. Phys. Lett., 8:116 (1966).
64. L. E. Erickson and A. Szabo, J. Appl. Phys., 37:4953 (1966).
65. L. E. Erickson and A. Szabo, J. Appl. Phys., 38:2540 (1967).
66. R. D. Maurer, Proc. Sympos. Optical Masers, New York (1963), p. 435.
67. A. G. Fox and T. Li, Bell System Tech. J., 40:453 (1961).
68. G. D. Boyd and J. P. Gordon, Bell System Tech. J., 40:489 (1961).
69. L. A. Vainshtein, Zh. Éksp. Teor. Fiz., 44:1050 (1963).
70. L. A. Vainshtein, Zh. Éksp. Teor. Fiz., 45:684 (1963).
71. D. A. Kleiman and P. P. Kislink, Bell System Tech. J., 41:453 (1962).
72. J. A. Fleck, J. Appl. Phys., 34:2997 (1963).
73. N. Kumagai, M. Matsuhara, and H. Mori, IEEE J. Quantum Electronics, QE-1:85 (1965).
74. N. S. Petrov and B. B. Boiko, Zh. Prikl. Spektrosk., 2:84 (1965).
75. B. B. McFarland, R. H. Hoskins, and B. H. Soffer, Nature, 207:1180 (1965).
76. B. J. McMurtry and A. E. Siegman, Appl. Opt., 1:51 (1962).
77. B. J. McMurtry, Appl. Opt., 2:767 (1962).
78. V. I. Malyshev, A. S. Markin, A. V. Masalov, and A. A. Sychev, FIAN Preprint, No. 3 (1969); Zh. Éksp. Teor. Fiz., 57:827 (1969).
79. M. Michon, J. Ernest, and R. Auffret, Phys. Lett., 21:514 (1966).
80. V. V. Korobkin and M. Ya. Shchelev, Zh. Éksp. Teor. Fiz., 53:1230 (1967).
81. L. Waszak, Proc. IEEE, 52:428 (1964).
82. V. I. Malyshev and A. S. Markin, Zh. Prikl. Spektrosk., 6:471 (1967).
83. W. R. Sooy, Appl. Phys. Lett., 7:36 (1965).
84. T. V. Gvaladze, I. K. Krasyuk, P. P. Pashinin, A. V. Prokhindeev, and A. M. Prokhorov, Zh. Éksp. Teor. Fiz., 48:106 (1965).
85. V. I. Malyshev and S. G. Rautian, Opt. i Spektrosk., 6:550 (1959).
86. E. Snitzer, Appl. Opt., 5:121 (1966).
87. P. P. Feofilov, A. M. Bonch-Bruevich, V. V. Vargin, et al., Izv. Akad. Nauk SSSR, Ser. Fiz., 27:466 (1963).
88. V. Daneu, C. A. Sacchi, and O. Svelto, IEEE J. Quantum Electronics, QE-2:290 (1966).
89. V. I. Malyshev, A. S. Markin, and A. A. Sychev, Zh. Prikl. Spektrosk., 10:248 (1969).
90. T. I. Kuznetsova, V. I. Malyshev, and A. S. Markin, Zh. Éksp. Teor. Fiz., 52:438 (1967).
91. E. L. Steele, IEEE J. Quantum Electronics, QE-1:42 (1965).
92. C. A. Sacchi, G. Soncini, and O. Svelto, Nuovo Cim., B-48:58 (1967).
93. A. Schmackpfeffer and H. Weber, Phys. Lett., 24A:190 (1967).
94. R. Marrach and G. Kachen, J. Appl. Phys., 39:2482 (1968).
95. A. Schawlow, in: Lasers [Russian translation], IL (1963), p. 54; see also: Proc. Conf. Optical Instruments and Techniques, Chapman and Hall, London (1962), pp. 431-440.
96. M. A. Duguay, S. L. Shapiro, and P. M. Rentzepis, Phys. Rev., Lett., 19:1014 (1967).
97. J. A. Giordmaine, P. M. Rentzepis, S. L. Shapiro, and K. W. Wecht, Appl. Phys. Lett., 11:216 (1967).

98. T. I. Kuznetsova, FIAN Preprint, No. 25 (1967).
99. V. S. Letokhov and V. N. Morozov, Zh. Éksp. Teor. Fiz., 52:1296 (1967).
100. E. M. Garmire and A. Yariv, IEEE J. Quantum Electronics, QE-3:222 (1967).
101. W. H. Glenn and M. J. Brientza, Appl. Phys. Lett., 10:221 (1967).
102. I. K. Krasyuk, P. P. Pashinin, and A. M. Prokhorov, ZhÉTF Pis. Red., 7:117 (1968).
103. S. D. Kaitmazov, I. K. Krasyuk, P. P. Pashinin, and A. M. Prokhorov, Dokl. Akad. Nauk SSSR, 180:1331 (1968).
104. J. A. Armstrong, Appl. Phys. Lett., 10:16 (1967).
105. P. M. Rentzepis and M. A. Duguay, Appl. Phys. Lett., 11:218 (1967).
106. V. I. Malyshev, A. S. Markin, A. V. Masalov, and A. A. Sychev, FIAN Preprint, No. 34 (1968); Zh. Prikl. Spektrosk., 11:655 (1969).

DETECTION AND INVESTIGATION OF STIMULATED EMISSION IN A PINCHED DISCHARGE*

V. M. Sutovskii

INTRODUCTION

There has been a noticeable trend in recent years toward an interfusion of the techniques of quantum electronics and high-temperature plasma physics. Powerful optical quantum oscillators (lasers) are widely used for both the generation and diagnostics of a plasma. On the other hand, plasma light sources (pinch discharges, exploding wires, etc.), with their high energy output in the visible and ultraviolet regions of the spectrum, are very well suited to the optical pumping of lasers.

An entirely new area of application of high-temperature plasmas in quantum electronics is the use of fast processes and various instabilities inherent in such plasmas for the stimulated emission of radiation. The idea of using nonequilibrium processes in a high-temperature plasma for the fabrication of lasers was proposed in the Accelerator Laboratory of the P. N. Lebedev Physics Institute. One aspect of this research was the attempt to realize lasing in a direct pinch discharge. Lasing in this type of discharge was first obtained in 1965 from argon ions Ar II [1] and later from doubly ionized argon [2]. Soon afterwards in the United States similar experiments were conducted with theta-pinch [3]. Of major importance is the study [4], in which lasing was obtained in a plasma-beam discharge. The beam instability caused by the interaction of high-energy electrons with the plasma is responsible for the creation of the active medium in this type of discharge.

The realization of negative-temperature states in a pinch discharge opens up new possibilities with respect to gas lasers. The high density of active centers and energy influx rate in the constricted filament of a powerful pinch discharge makes it possible to obtain very large pulsed radiation powers and, in principle, to advance into the short-wave part of the spectrum, including the vacuum ultraviolet region.

The observation of stimulated emission in a high-temperature plasma is also of fundamental significance with regard to plasma physics. First of all, the detection of emission is evidence of strong deviations from thermodynamic equilibrium. The investigation of the emission characteristics can provide additional information about the nonequilibrium processes in

---

* Based on the author's dissertation presented June 17, 1968, under the direction of Prof. M. S. Rabinovich and V. M. Likhachev at the P. N. Lebedev Physics Institute of the Academy of Sciences of the USSR.

a plasma. Second, stimulated emission can powerfully affect the configuration of the spectral lines and total amount of energy lost by the plasma. In calculating the radiation losses, therefore, it is required to take account of stimulated emission [5]. This requirement is particularly essential in application to the large plasma formations typical of cosmic objects.

The recent vigorous development of quantum electronics has greatly expanded the types of gas lasers and their capabilities. The number of transitions in which lasing is obtained now exceeds 600, and the emission wavelength interval has been extended from the far infrared into the ultraviolet region (0.1 mm to 2300 Å) of the spectrum. A large fraction of the laser transitions in the visible and ultraviolet spectra are attributable to ions of various elements.

Although the nature of the excitation of ionic lasers is by no means completely understood at this time, it seems reasonable that the realization of a sufficiently large gain requires the active medium to contain a large number of active centers (ions) and to have a high excitation rate. The latter requirement, in turn, refers to the short-wave region of the spectrum, due to the reduction of the energy-level lifetimes and gain with the wavelength. Typical field strengths (several kV/m) and current densities (hundreds of A/cm^2) for pulsed ionic lasers, even though they exceed the corresponding variables for ordinary gas lasers, do not, in our opinion, come up to the ultimately attainable capabilities of ionic lasers. It would be interesting, therefore, to use as the active medium of an ionic laser a high-temperature plasma from powerful pulsed discharges with parameters several orders of magnitude greater than usual.

Of the possible types of pulsed discharges we chose for the present investigation the high-current self-constricted discharge, or pinch, widely used in high-temperature plasma physics. A distinctive attribute of this type of discharge, besides its high temperature and degree of ionization, is its strong deviation from the thermodynamic equilibrium state due to the exceedingly fast transient process, rapid heating, and the development of instabilities (beam and others) at the instant of cumulative discharge. Inasmuch as the characteristic times of these processes are commensurate with the lifetimes of the upper laser levels, it is logical to assume that negative-temperature states must exist in a pinch-discharge plasma.

The present study is given over to an investigation of stimulated emission in a direct pinch-discharge plasma. In the first chapter we briefly review the work to date on pulsed ionic lasers. The second chapter is devoted to a description of the physical processes involved in a pinch discharge. The possibilities for obtaining negative-temperature states and the experimental plan are discussed briefly. In the third chapter we describe the experimental apparatus and give the current, optical, and spectral discharge characteristics. In the fourth chapter we describe the procedure and results of an investigation of stimulated emission by argon ions with varying ionization multiplicities. In the fifth and final chapter we give the results of measurements of the parameters of a discharge plasma, whereupon we are able to formulate a qualitative interpretation of the results described in the fourth chapter. We examine the possibility of diagnostics of a plasma on the basis of scattering of its own stimulated emission. We conclude with a discussion of the experimental results and the outlook for the application of pinch discharges for lasing in the vacuum ultraviolet region.

CHAPTER I

STATE OF THE ART OF PULSED IONIC LASER RESEARCH

The multivariety of gas lasers can be classified, irrespective of their specific type of active medium, into two major groups according to their operating regime: pulsed and continuous lasers. Logically enough, the class of pulsed laser transitions is more numerous. The necessary conditions for continuous or pulsed lasing can be determined from the notions concerning local thermodynamic equilibrium (LTE). For simplicity we shall neglect all excitation processes except the electronic type. Thus, radiative excitation can be entirely neglected due to the small optical thickness of the medium in gas lasers. If we confine our perspective solely to one-component systems, we thereby eliminate resonance excitation processes associated with atomic collisions, which can contribute to the excitation of the laser levels in multicomponent systems. The converse processes of de-excitation are determined both by electron collisions of the second kind and by spontaneous emission. Triple and radiative recombination compete in the process of recombination. The relative contribution of collision and radiative processes to the population balance of individual levels and to ionization equilibration depends on the electron density. We can distinguish three cases, depending on the intensity of radiative and collision processes.

1. LTE Model. The electron density is so large that the role of collisions of the second kind and triple recombination prevails. The populations of the individual levels, as in the case of total thermodynamic equilibrium, fit a Boltzmann distribution.

2. Collision–Radiation Model. The role of collisional and radiative processes is comparable in the depopulation of individual levels and in recombination. In this case, however, a Boltzmann distribution is possible between the individual levels. This statement refers to the upper levels, as such a distribution does not occur between lower levels (see Chapter V for further details).

3. Corona Model. The electron density is so small that the role of collisions of the second kind and triple recombination is insignificant by comparison with radiative processes. In this case the populations of the individual levels are determined by the balance between electron collisions and spontaneous emission.

It is obvious that continuous lasing is only possible if the corona model is applicable or between lower levels in the case of the collision–radiation model. The possibilities for pulsed lasing are far richer. We now enumerate some of the principal advantages of the pulsed stimulation of emission:

a. In the pulsed regime one can obtain a large driving power, which is technically unfeasible in the continuous regime.
b. In the presence of powerful excitation it is possible to create a transient inversion at large electron densities up to the values at which LTE exists under steady-state conditions.
c. Pulsed excitation makes it possible to create an inversion between levels such that the lifetime of the upper level is shorter than that of the lower, permitting the range of laser transitions to be significantly expanded. In this case the selection rules are such that only the upper level is dominantly excited, so that very large population inversions become possible. However, the lasing period in this case is limited by the lifetime of the upper laser level.
d. In the pulsed regime lasing is possible at transitions for which the gain in the continuous regime is inadequate to overcome the losses in the cavity or to stimulate emission.

In addition to the classification of lasers according to their emission regime, we can also classify them according to the type of active medium: molecular, atomic, and ionic lasers. The diversity of characteristics of these lasers is attributable to the differences in the energy level systems. The latter accounts primarily for the difference in the spectral intervals of emission. Thus, molecular systems are characterized by the closest-spaced groups of levels, between which the transition energy is ≤ 0.1 eV, which corresponds to the far infrared region of the spectrum. As far as atomic and ionic lasers are concerned, deviations from partial LTE are possible for not too high levels. The energy of the corresponding atomic transitions is 1 or 2 eV and corresponds to the visible region. The energy of the analogous ion transitions is significantly greater, according to the Bohr frequency rule:

$$h\nu \sim Z^2 \left(\frac{1}{n_0} - \frac{1}{n}\right)^2,$$

i.e., with increasing ionization multiplicity the transition energy increases as the square of the charge of the ion sheath. Ionic laser transitions are therefore generally concentrated in the ultraviolet region of the spectrum.

However, the creation of population inversion in transition to adjacent systems of ion levels is difficult on account of the reduction in lifetime of the corresponding levels:

$$\tau = \frac{3cm}{8\pi^2 e^2} \lambda^2,$$

where m and e are the electron mass and charge, c is the velocity of light, and λ is the laser transition wavelength. The gain of the active medium and rate of excitation of the upper laser level in the case of Doppler broadening of the spectral lines are related by the expression

$$G \sim F\lambda^3,$$

in which G is the gain, F is the excitation rate, and λ is the laser transition wavelength. Thus, with a decrease in the wavelength the requirements on the excitation intensity are increased. However, the pulse power in this case must increase due to the reduction in de-excitation time and increase in the photon energy.

These characteristics of ionic lasers as a whole have motivated a tremendous surge of activity in recent years. Emission has now been realized with all the inert gases except helium and radon, as well as with most of the elements in the second, fifth, sixth, and seventh groups of the periodic table. The number of ionic laser transitions has reached several hundred, and they span the entire visible, ultraviolet, and part of the infrared regions of the spectrum. We do not propose in the present article to give a detailed survey of research on pulsed ionic lasers. We shall merely highlight the principal landmarks in the development of this topic, delving in somewhat more detail into argon lasers and lasers operating purely on plasma devices.

The first pulsed emission in ionized-atom transitions was obtained toward the end of 1963 by Bell in the United States [6, 7] and by Armand and Martinot-Lagarde in France [8]. In these studies lasing was observed in a mixture of helium and mercury vapor in the visible and ultraviolet regions. Emission was obtained in the afterglow in tubes of small diameter (3 to 15 mm) and 3 m in length at currents up to 50 A. The emitted pulse width was several microseconds.

The observation of pulsed emission from argon ions was first reported by Bridges [9], W. R. Bennett et al. [10], and Convert et al. [11] soon after the realization of emission from mercury ions. The gain for argon ions is so great that lasing was sometimes observed in a mercury laser containing a minute argon impurity. In [9] emission was obtained at ten Ar II lines in the interval from 4545 to 5287 Å. The discharge was produced in a quartz tube 4 mm in diameter and about a meter in length. The current pulses had a width of 0.5 to 7 μsec and an amplitude of 40 A. The peak power of the emission was only 1 W. We should consider [10] somewhat more in detail. In this study W. R. Bennett and co-workers observed emission at several Ar II transitions without mirrors, i.e., superradiance. It was the authors' opinion that such a large gain is caused by the large values of E/p, equal to 1000 W/cm • torr in the cited investigation. Emission was observed in tubes with an inside diameter of 5 mm and length of 1 m, using the discharge of a 0.001-μF capacitor at a voltage up to 20 kV. Under these conditions the lasing period was 20 nsec, and the power attained several hundred watts. The optimum pressure was 0.02 torr. Quasi-continuous lasing for a period of about 1 msec at 10 W was observed in the same study with the use of another power supply (C = 100 μF, v = 6 kV).

After the discovery of emission from argon and mercury ions lasing was realized with the ions of several other elements: noble gas ions and elements of the second, fifth, sixth, and seventh groups of the periodic table. A great many laser lines have been obtained in an induction-excited ring discharge [12]. The advantage of the ring discharge is the absence of electrodes, so that it is possible to work with active elements that normally react with electrode materials. Moreover, the ring discharge has the virtue that the system can be heated to the very high temperatures needed in connection with metallic vapors [13]. A survey of research on ionic lasers may be found in the literature [14-16]. We merely point out that, despite the tremendous diversity of ionic lasers, they all have several properties in common. A great many emission lines are obtained in small-diameter (1-7 mm) tubes. The tube length ranges from 0.5 to 3 m. The discharge currents do not usually exceed 1 or 2 kA. Much of the work done so far has been of a piecemeal nature, and the mechanisms for the excitation of the laser levels have remained virtually uninvestigated. An exception is the case of the argon ionic laser, for which the lifetimes and cross sections of electronic excitation have been measured. These data allow one to assess the role of various processes in the filling of the laser levels. The lifetimes of the laser levels in argon were first measured by W. R. Bennett and co-workers [17]. Similar measurements were later performed in [18] (see also [19]). The results of the measurements in both of these papers exhibit close agreement. The transition probabilities and lifetimes of the individual levels for Ar II have also been calculated theoretically in [20-22]. The calculations were carried out on the assumption of pure LS-coupling with the use of the wave functions calculated by the Hartree–Fok method. The good agreement of the experimental and theoretical data is noteworthy. The contribution of cascade filling of the laser levels has been estimated in [23-27]. In [27] the most complete data are given on the lifetimes and transition probabilities of a large number of Ar II levels. Many of them were calculated for the first time. In the aggregate these data enabled the authors to determine the populations of the individual levels and the gain for a number of laser transitions at various pumping levels. Despite the fact that the data refer to conditions incumbent in continuously emitting argon lasers, they nevertheless permit the required estimates to be made for pulsed lasers as well, using the cross sections given in [28, 29] for direct excitation of the upper laser levels from ground-state neutrals. The argon ionization cross sections needed for calculation of the cascade excitation are given in [30]. The configuration of the Ar II laser levels is illustrated in Fig. 1. The corresponding lifetimes and transition probabilities, taken from the cited papers, are listed in Table 1.

Since about 1965 a number of papers have been published in which high-current pulsed discharges and other systems invested with strongly pronounced nonequilibrium processes were used for the excitation of laser levels. We give special attention to those papers in which lasing was obtained in direct- and theta-pinch discharges [1-3, 31, 32]. The present article is concerned with the investigation of lasing in a direct pinch. We shall therefore confine our discussion to [3]. Emission is obtained in θ-pinch at two transitions in ionized nitrogen: 4321 and 4329 Å. The laser pulse consists of several peaks 0.2 μsec in width each, with a pulse power of 40 kW. The energy parameters of the discharge are similar to those described in [1, 2]. The energy source was a 0.5-μF capacitor at voltages from 15 to 27 kV. Beyond these limits lasing was not observed. Discharge was realized in short chambers 10 cm in length and 2.5 cm in diameter at a working-gas pressure of ~ 1 torr. Another study deserving detailed consideration is [4]. In it the role of plasma instability (beam type) in the creation of the active medium is

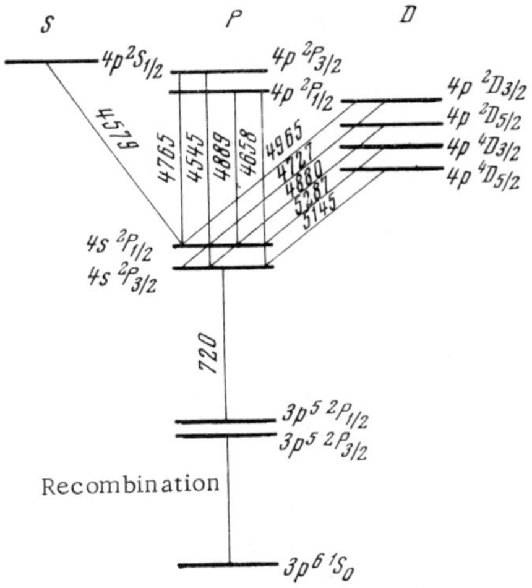

Fig. 1. Laser-level diagram for Ar II.

TABLE 1. Lifetimes and Transition Probabilities for Selected Laser Levels of Ar II

Upper lasing level	Transition	Emission wavelength, Å	Lifetime of upper laser level, 10^{-9} sec				Laser transition probability, 10^{-7} sec^{-1} [27, 21]
			experiment		theory		
			[17]	[18]	[27]	[21]	
$4p^2P^0_{3/2}$	$4p^2P^0_{3/2} - 4s^2P_{1/2}$	4764.89	9.5±0.5	8.5±0,14	9.27	7.9	4.47
$4p^2D^0_{5/2}$	$4s^2D^0_{5/2} - 4s^2P_{3/2}$	4879.9	9,1±0.6	10.3±0,2	8.52	9.5	8.45
$4p^2D^0_{3/2}$	$4p^2D^0_{3/2} - 4s^2P_{3/2}$	4726.8	9.8±0,2	—	9,27	9.7	4.56
$4p^2P^0_{1/2}$	$4p^2P^0_{1/2} - 4s^2P_{3/2}$	4657,9	8.7±0,3	8.0±0.13	8.85	8,7	8.5
$4p^2S^0_{1/2}$	$4p^2S^0_{1/2} - 4s^2P_{1/2}$	4579.3	8.8±0.3	8.7±0.15	7.32	6 2	9.0

directly demonstrated. Lasing was obtained from Ar II ions in a plasma-beam discharge. Discharge was produced in a chamber 100 cm in length and 20 mm in diameter with the pulsed transmission of a beam of high-energy electrons (on the order of several kiloelectron-volts) at a beam current up to 40 A. Beam instability occurred upon interaction of the electron beam with the plasma. A medium was thus created with a high electron temperature (100 eV) and a relatively low ion temperature (1 eV). Lasing occurred at many Ar II transitions and had a duration of 30 μsec and power of 100 W. It is important to note the totally unexpected absence of lasing at the 1764.89 Å line, which has the maximum intensity in the majority of high-current pulsed lasers.

The use of a transverse discharge with a traveling excitation wave is intriguing [33-35]. In this system Kolb and Shipmen [35] obtained very short current pulses 20 nsec in width and up to 500 kA in amplitude. The current rise had a width of just a few nanoseconds. An extremely high lasing power was obtained in molecular nitrogen and in neon using this type of discharge. Powers of 2.5 and 0.7 MW were obtained in nitrogen and neon, respectively. Clearly, it should also be possible in this kind of system to obtain high-power lasing at the various ion transitions at which emission takes place in conventional lasers.

With regard to the excitation mechanisms for pulsed ionic lasers, we shall limit our discussion to electronic excitation. Electron-impact excitation is the most common approach to the problem of filling the upper laser levels and is responsible for the creation of population inversion in the majority of gas lasers. The possibility of the electronic excitation of inversion was first indicated in [36-38]. The specific excitation mechanisms involved in each case have, of course, their own unique characteristics and require special consideration. The general requirements imposed on the characteristics of the active medium and the concomitant properties of the emission were described at the beginning of the chapter. We now look at some specific applications.

In the event that the state of the system is characterized by the temperature, i.e., if it is in equilibrium, the only way to effect inversion with the application of electron collisions is by a rapid change in temperature of the system. This method has been investigated by Basov and Oraevskii [39]. They showed that population inversion can occur in the relaxation process associated with rapid heating or cooling in a three-level system. We assume that the relaxation processes in a three-level system are characterized by the transition probabilities w_{ij} per unit time from the i-th to the j-th level. Given a definite relation between the probabilities of transitions between levels 1, 2, and 3 (numbered in order of increasing energy), emission becomes possible. Thus, let $w_{13} \gg w_{12}, w_{32}$. This means that with rapid heating of the system in a time $t_h < 1/w_{13}$ equilibrium will develop rapidly between levels 1 and 3, whereas between levels 1 and 2 such equilibrium will be lacking for a certain period of time. Consequently, a state will set in with population inversion between levels 3 and 2. To obtain more precise quantitative relations we assume that the three-level system is rapidly heated from an initial temperature T_s to a certain final temperature $T_f > T_s$. We denote the number of particles at each level by n_1, n_2, and n_3. These quantities are functions of the time and satisfy the equation

$$n_1 + n_2 + n_3 = \text{const.} \tag{1}$$

The kinetic equations for all n are written in the form

$$\begin{aligned} \dot{n}_3 &= -(w_{31} + w_{32})\, n_3 + w_{32} n_2 + w_{13} n_1, \\ \dot{n}_2 &= -(w_{21} + w_{23})\, n_2 + w_{32} n_3 + w_{12} n_1, \\ \dot{n}_1 &= -\dot{n}_2 - \dot{n}_3. \end{aligned} \tag{2}$$

The relaxation processes have an aperiodic character. The qualitative form of the possible time variations of the populations is shown in Fig. 2. Following is a necessary and sufficient condition for population inversion:

$$\max \Delta n = \max (n_3 - n_2) = A\,(1 - a)\, e^{-\lambda_1 t_m} + B\,(1 - b)\, e^{-\lambda_2 t_m} + n_{30} - n_{20} > 0, \tag{3}$$

in which

$$t_m = \frac{1}{\lambda_2 - \lambda_1} \ln \left[-\frac{B\,(1-b)}{A\,(1-a)} \frac{\lambda_2}{\lambda_1} \right] > 0, \tag{4}$$

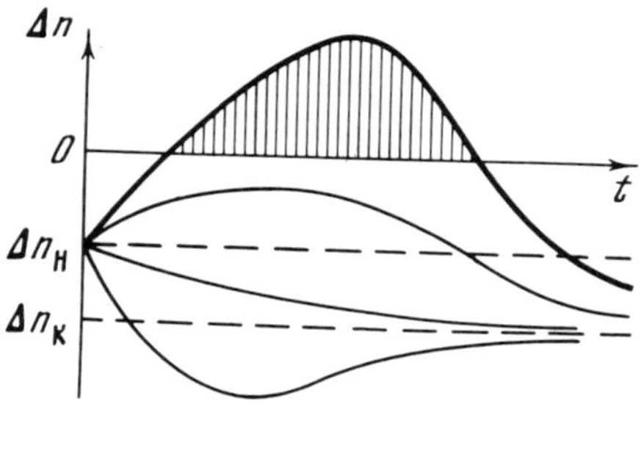

Fig. 2. Variation of population difference between levels 3 and 2 for rapid heating of a three-level system.

where

$$A = \frac{n_{20} - N_2 - (n_{30} - N_3)\,b}{b - a}, \qquad B = \frac{n_{20} - N_2 - (n_{30} - N_3)\,a}{a - b};$$

N_3, N_2, n_{30}, and n_{20} are the equilibrium populations at the initial and final temperatures, respectively; $a = (\lambda_1 - \alpha)/\beta$; $b = (\lambda_2 - \alpha)/\beta$; $\alpha = w_{31} + w_{13} + w_{32}$; $\beta = w_{13} - w_{23}$; and λ_1 and λ_2 are the roots of the characteristic equation (3) of the system. Conditions (3) and (4) give the required relations between T_s and T_f and the transition probabilities w_{ij}. In particular, for $T_s \ll h\nu_2/k$ we have $N_2 \approx N_3 \approx 0$, and inequalities (3) and (4) are equivalent to the simple relation

$$w_{12} < w_{13}. \tag{5}$$

It follows from this inequality that, due to the rapid filling of level 3 as compared with level 2, population inversion exists between them for a certain period of time. Analogous calculations can be carried out for systems having more levels, but the analysis in this case is far more complex, although nothing fundamentally new is learned over the three-level system. Similar effects must also take place in rapid cooling of the system, as, for example, in recombination. This method has been elaborated by Shelepin and Gudzenko (see, e.g., [40]).

In each specific instance the difference in the excitation probabilities is due to different factors. In the majority of ionic lasers the selection rules due to instantaneous excitation are used. The method of instantaneous perturbations was first used by Lamb [41] to calculate the excitation probabilities for the states of He II. The method characterizes the class of ion levels excited directly from ground-state neutrals. It is possible, therefore, to determine which of the following two processes takes place from the nature of the upper laser levels:

Direct excitation from the ground state of a neutral:

$$e + X_{\mathrm{gr}} \to X^{+}{}_{\mathrm{ex}} + 2e. \tag{6}$$

Step-by-step excitation, which can be realized, for example, by two processes in succession:

$$\begin{aligned} e + X &\to 2e + X^{+}{}_{\mathrm{gr}} \\ e + X^{+}_{\mathrm{gr}} &\to e + X^{+}{}_{\mathrm{ex}} \end{aligned} \tag{7}$$

Other step-by-step excitation processes are possible. For instance, a metastable state of a neutral can become filled first, then as a result of a second collision an excited ion state is filled. Also, ion levels can be excited above the upper laser level, which in this event becomes filled as a result of cascade transitions.

The instantaneous perturbation method, which predicts the class of levels excited most intensely, makes it possible to invoke certain information on the nature of the excitation. W. R. Bennett used this method to account for the distribution of excited ion levels in inert gases under α-particle excitation [42], and then to interpret the action of a pulsed argon-ion laser in high-current low-pressure discharges (see, e.g., [16]). We now describe the method according to [16].

It can be shown on the basis of nonstationary perturbation theory that the probability P_n that the system will remain in the n-th perturbed eigenstate after the perturbation potential has been turned on or off "infinitely fast" is given by the expression

$$P_n = |a_n|^2, \tag{8}$$

where

$$U_0 = \Sigma a_n \psi_n.$$

Here U_0 is the initial wave function for the system, and ψ_n represents the proper wave functions of the new Hamiltonian after instant variation of the potential. The expression "infinitely fast" connotes that the energy variation ΔE of the system occurs in a time $t \ll \hbar/\Delta E$. For fast ionizing collisions of the type

$$e + X \rightarrow (X^+)^* + 2e,$$

where $(X^+)^*$ is an excited ion state, this type of consideration means that the wave function for the system of electrons left in the shell cannot change during the collision period. The validity of the method depends, of course, on the energy of the primary electron. Assuming the excitation cross section is $\sim 10^{-18}$ cm^2, we find that the instantaneity condition $t \ll \hbar/\Delta E$ is satisfied for an electron energy

$$E \gg \left(\frac{\Delta E}{\hbar}\right)^2 \sigma m \approx (\Delta E)^2 \cdot 10^9.$$

The method affords a reasonable approximation for energies several electron-volts above the threshold value. Quantum-mechanical analysis shows that for the most part states will be excited having a configuration similar to that of the shell immediately after the emission of an electron. In many cases this situation leads to the predominant formation of ion levels having the same parity as the ground ion state. This makes it possible in accordance with the selection rules for dipole radiation to use the radiative depopulation of the lower laser level due to resonance transitions in the deep ultraviolet region of the spectrum (see Fig. 1). With step-by-step excitation, on the other hand, mainly states opposite in parity to the ground ion state are excited. Different gains are obtained at the same transitions in different experiments, depending on the nature of the excitation (direct or step-by-step). The assumption of pure LS-coupling and instantaneous excitation for argon gives predominant excitation of the laser transition with wavelength 4764.89 Å ($4p^2P^0_{3/2} \rightarrow 4s^2P_{1/2}$). The same inference is also drawn from an experimental measurement of the direct-excitation cross sections for various levels from the argon neutral ground state [28]. Thus, W. R. Bennett and co-workers [10] observed the maximum gain at the wavelength 4764.89 Å. The other transitions observed in argon under these conditions belong to the same configuration and can be excited if LS-coupling is violated or if there are deviations from instantaneous excitation. The dominant role of instantaneous excitation processes is further supported by the results of experiments on the stimulation of emission by an electron beam [43]. The most intense lasing in this investigation was obtained at the 4764.89 Å line. Moreover, the lasing intensity depended linearly on the argon density and on the current, a situation that also corresponds to direct excitation. By contrast, in continuous lasers operating at relatively low currents the number of high-energy electrons is small, and the laser levels must be excited by cascade processes. In this case the maximum gain is observed at the transition with wavelength 4879.9 Å [44]. The cascade mechanism is corroborated by the square-law dependence of the lasing intensity on the current. The maximum gain at the 4879.9 Å line in the pulsed regime as obtained by Bridges [9] is unexpected. W. R. Bennett attributes this fact to the use of an oxide cathode in Bridges' work and of a cold cathode in all papers in which the maximum gain was observed at the 4764.89 Å line. Further evidence of the prevalence of instantaneous excitation processes in ionic lasers is the fact that a large part of the laser transitions begin with a p-configuration, in accordance with the predictions of instantaneous excitation theory.

Despite the large role of electron collisions in the stimulation of emission, they clearly are not always the only lasing mechanism. Thus, in [45, 46] lasing was observed from argon ions not only at the beginning of the current pulse, but also at its end or in the afterglow, when there are no longer any high-energy electrons. The most probable mechanism in this case is the creation of inversion as a result of recombination and optical cascade transitions. The difference between the excitation mechanisms at the beginning and end of the current pulse is also evinced by the different spectral compositions of the emission lines. At the beginning of the current pulse the 4764.89 Å line is the most intense, whereas in the afterglow only the 4879.9 Å line is left.

At the large current densities typical of ionic lasers plasma interactions caused by the high degree of ionization and high temperature play a vital role in the excitation processes. Plasma interactions result in the development of instabilities, acceleration processes, and local temperature and density jumps. All of these events have the aggregate effect of increasing the fraction of high-energy electrons and increasing the intensity of the excitation processes. A correlation between the thermal noise generated by collective interactions in the plasma and laser emission has been observed even at small currents in the operation of a helium–neon laser [47, 48]. The role of plasma interactions in the operation of ionic lasers has been indicated by W. R. Bennett [10]. In the latter study the lasing intensity exhibited fluctuations coinciding timewise with the discharge current fluctuations. The role of plasma instabilities is very clearly pronounced in an argon laser driven by a plasma-beam discharge [4]. In our opinion, however, pinch discharges are characterized by the greatest diversity of plasma effects, which are of independent interest, as well as holding definite promise with regard to practical applications.

CHAPTER II

ON THE CREATION OF NEGATIVE-TEMPERATURE STATES IN A PINCH-DISCHARGE PLASMA

The self-constriction of an electric current by virtue of its own magnetic field (the pinch effect) has been known for some time. It was first investigated by W. H. Bennett (Sr.) for a relativistic electron beam [49] (1934).

There was a growing interest in discharges after World War II in connection with the work being done on controlled thermonuclear reactions. The rapid pinch of a discharge makes it possible to attain very high temperatures, but the development of plasma instabilities leads to breakdown of the plasma column and lowers the attainable upper temperature limit.

From the standpoint of quantum electronics we are interested both in the pinch effect *per se* and in the subsequent development of instabilities. These processes cause exceedingly large deviations from the thermodynamic equilibrium state in the pinch-discharge plasma and can result in the formation of negative-temperature states.

§1. Physical Processes in a Pinch Discharge

The pinch discharge differs from the more conventional discharges used for the creation of pulsed ionic lasers by the high values of its discharge current buildup rate (10^{10} to 10^{12} A/sec) and the large values of its current at the instant of cumulative discharge (up to 10^6 A). The pinch effect is usually realized in cylindrical chambers up to 10 or 20 cm in diameter and normally no more than 1 m in length. The two most common types of pinch are the direct, or zeta-pinch

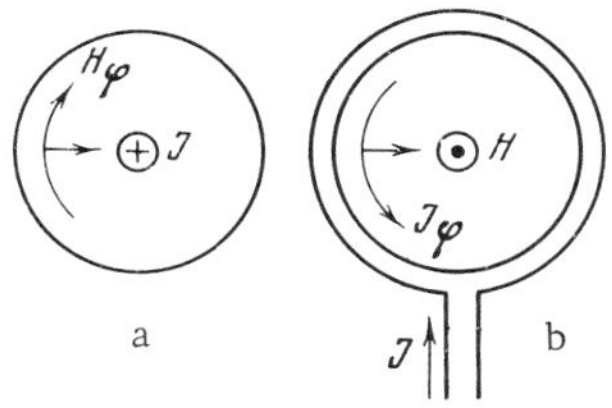

Fig. 3. Z-Pinch (a) and θ-pinch (b) schemes.

(Z-pinch) and the theta-pinch (θ-pinch) (Fig. 3). Zeta-pinch takes place in a straight cylindrical tube with electrodes at the ends. The discharge is pinched under the action of an azimuthal magnetic field. Theta-pinch is an electrodeless discharge. It is induced by current flowing through coils enclosing the discharge chamber. Consequently, θ-pinch is also called induction pinch. The main physical processes in both types of discharge are similar. We analyze their action in the example of the Z-pinch investigated in the present article.

The formation and development of pinch can be broken down into the following stages:

1. electrical breakdown of the medium (avalanche formation, weak ionization of the gas);
2. expulsion of the magnetic field from the central part of the discharge (skin effect) and magnetization of electrons;
3. constriction of the plasma sheath by electrodynamic forces and the formation of a plasma filament;
4. rapid heating of the plasma at the instant of constriction of the discharge to the axis of the chamber (cumulative discharge effect);
5. deflection and disruption of the plasma filament due to the development of plasma instabilities.

The discharge develops so rapidly that strong deviations from the thermodynamic equilibrium state can persist throughout the entire discharge. We now examine each stage of the discharge process in closer detail. We rely mainly on [50-54].

1. Breakdown. By breakdown we refer to the effects taking place in the gas prior to the onset of internal electric and magnetic fields. In this state, therefore, the degree of ionization and discharge current are slight. The resistance of the discharge gap is greater than its reactance. At the low initial pressures in a pinch discharge (10^{-3} to 1 torr) the breakdown is not localized, but is spread over the entire cross section of the discharge chamber. The electron temperature in breakdown is higher than the ion and atomic temperatures. For not too large E/p, however (E is the electric field strength, and p is the initial pressure) the electron temperature remains small (~ 3 eV) due to excitation and ionization losses. The ionization process is described by the following equation in the breakdown phase:

$$\frac{\partial n}{\partial t} = \nu n + D\nabla^2 n + \mu E \nabla n, \tag{1}$$

where n is the electron density; ν is the ionization number, or number of ionization events per electron per second; and μ and D are the mobility and diffusion coefficients of electrons. The mobility and diffusion of ions, due to the large mass of the latter, can be neglected. The diffusion coefficient is equal to

$$D = \frac{T_e \tau}{m},$$

and the mobility μ is related to D by the Einstein equation

$$\mu = \frac{eD}{T_e} = \frac{\tau e}{m},$$

in which T_e, m, and e are the temperature, mass, and charge of the electron and τ is the electron mean-free transit time.

With a change in the electron density the conductivity of the plasma also changes in accordance with the relation

$$\sigma = \frac{e^2 n\tau}{m}, \tag{2}$$

so that

$$\frac{d\sigma}{dt} = \frac{e^2\tau}{m}\frac{dn}{dt}. \tag{3}$$

Hereinafter we shall consider τ to be time-independent. This assumption does not inject an appreciable error into our qualitative analysis, because the electron temperature changes only slightly in the initial discharge phase. For small electron densities, such that Coulomb interactions can be neglected, the mean free transit time is determined by the cross section for elastic scattering by neutrals. Data on these cross sections may be found in [55]. The value of τ for argon at $T_e \sim 3$ eV and for an initial density of $3 \cdot 10^{14}$ cm^{-3} is 10^{-8} sec. The ionization frequency is

$$\nu = \eta E \bar{u},$$

where η is the ionization coefficient, or number of ion pairs created by a single electron across a potential difference of 1 V, and $\bar{u}$ is the mean velocity of ordered electron motion. This velocity is related to the mean velocity $\bar{v}$ of random motion as follows:

$$\frac{\bar{v}}{\bar{u}} \sim \sqrt{\frac{M}{4m}},$$

where M is the mass of the atom or ion. For argon at $T_e = 3$ eV we find $\bar{u} = 10^6$ cm/sec. The coefficient η is given as a function of E/p for argon in [56]. Under our experimental conditions (E = 100 V/cm, p = 10^{-3} torr) we find that $\nu \gtrsim 10^7$ sec^{-1}. Neglecting the mobility and diffusion, we find the conductivity growth law:

$$\frac{d\sigma}{dt} = \nu\sigma = \frac{\nu e^2\tau}{m} n \tag{4}$$

or, under our experimental conditions,

$$\frac{d\sigma}{dt} = 2.5 \cdot 10^7 n.$$

The development of discharge is determined by the avalanche buildup process as long as the plasma conductivity is fairly small, i.e., as long as the resistance of the plasma is considerably greater than the reactance of the discharge circuit.

With an increase in the conductivity, $\sigma > 1/\omega L$, the ionization buildup processes are determined by the electrical parameters of the circuit. The limiting value of the electron density is found from the condition

$$\frac{1}{\sigma}\frac{l}{s} = \omega L,$$

where l is the length of the discharge chamber, s is the cross section of the chamber, $\omega = 2\pi/\sqrt{LC}$ is the natural frequency of the discharge loop, L is the circuit inductance, and C is the capacitance. The value of n for which the foregoing discussion remains valid is equal to

$$n = \frac{m}{e^2\tau} \cdot \frac{l}{s\omega L} \cdot 9 \cdot 10^{11} \text{ cm}^{-3}. \tag{5}$$

2. Skin Effect and Magnetization. As the ionization and conductivity become larger the plasma acquires a very large current and internal electric and magnetic fields. The influence of the magnetic field is felt in two ways. First, the electric field is driven away from the discharge interior, creating the so-called skin effect. Second, the presence of the magnetic field diminishes the conductivity due to the magnetization of electrons. This effect has to be the strongest at the periphery of the discharge, because that is where the magnetic field is a maximum. As a result, the discharge can build up in the center of the chamber. Thus, depending on which effect prevails, the discharge builds up at the chamber walls and is then constricted by the internal magnetic field (this is the pinch effect in the usual sense of the word), or it develops at once in the center of the discharge chamber.

The magnetization of the plasma and the relative intensity of this process in comparison with the skin effect has been analyzed in [48]. Magnetization sets in in fields such that $\omega\tau \gg 1$, where $\omega = eE/mc$ is the cyclotron frequency. Also given in the same paper is a criterion by which it is possible to determine which of the two effects prevails:

$$k = 2\sqrt{\pi}\sqrt{\frac{l_i}{R}\frac{e^2}{mc^2}}\frac{\sqrt{n_0\tau E}\sqrt{\xi e}}{\sqrt{mM\nu}}, \tag{6}$$

where l_i is the ion mean-free path, which is determined mainly by charge exchange process; n_0 is the initial density; and ξ is the degree of ionization. The pinch discharge develops for values of $k < 1$.

A characteristic feature of the skin effect in a discharge is the fact that it can occur in a practically constant electric field due to the rapid conductivity variation. Thus, from the Maxwell equations we have

$$\begin{gathered}\nabla^2 E = \frac{4\pi}{c^2}\frac{dj}{dt} = 0,\\ \frac{dj}{dt} = \frac{d(\sigma E)}{dt} \approx E\frac{d\sigma}{dt} = \nu\sigma E,\end{gathered} \tag{7}$$

or, in cylindrical coordinates,

$$\frac{d^2E}{dr^2} + \frac{1}{r}\frac{dE}{dr} - \frac{4\pi}{c^2}\nu\sigma E_z = 0. \tag{8}$$

The solution of this equation is a zero-order Bessel function of an imaginary argument:

$$E_z = E_z^0 I_0\left(\sqrt{\frac{4\pi\nu\sigma}{c^2}}\,r\right).$$

The quantity

$$\delta = \sqrt{\frac{c^2}{4\pi\nu\sigma}} \tag{9}$$

characterizes the displacement of the field from the center and represents the thickness of the skin layer. It differs from the usual criterion by the substitution of ν for the field frequency. Under our experimental conditions the skin thickness in the initial stage of discharge buildup, while the resistance of the plasma is still large ($n \leq 10^{12}$ cm^{-3}), is $\leqslant 2$ cm, i.e., is commensurate with the diameter of the discharge chamber (diameter ≤ 4 cm). As the conductivity increases the skin thickness decreases. The conductivity increases in accordance with (3) until Coulomb collisions become prevalent. The corresponding limiting value of the electron density is found from the condition that the collision frequencies due to elastic scattering by neutrals and to Coulomb collisions are equal [57]:

$$\frac{1}{\tau} = \sigma_0 n_0 \sqrt{\frac{kT}{m}} = \frac{4\sqrt{2\pi}}{3} \frac{N_e e^4}{(kT_e)^2} \left(\frac{kT_e}{m}\right)^{1/2} \Lambda, \tag{10}$$

where σ_0 is the elastic scattering cross section and Λ is the Coulomb logarithm. The corresponding degree of ionization is

$$\xi = \frac{3(kT_e)^2 \sigma_0}{4\sqrt{2\pi} e^4 \Lambda}. \tag{11}$$

For argon at $T_e = 3$ eV and $\Lambda = 10$ we have $\xi = 5\%$. For an initial density $n_0 = 3 \cdot 10^{14}$ cm^{-3} this value corresponds to $N_e = 1 \cdot 10^{13}$ cm^{-3}. For a high degree of ionization the conductivity has a weaker dependence on the electron density and in the limit is determined only by the electron temperature:

$$\sigma = 1.3 \cdot 10^{13} T^{3/2} \text{ eV} \quad (Z = 1,\ \Lambda = 10). \tag{12}$$

We now consider the development of discharge at times when the electrical engineering parameters of the network are the controlling factors. In this case the ionization process can be described as follows:

$$\frac{d(ns)}{dt} = \frac{J^2 \eta}{\sigma s}, \tag{13}$$

where J is the discharge current. This expression gives the increase of the conductivity in the interval of electron densities from 10^{12} to 10^{13} in application to our experimental conditions. Substituting Eq. (2) into (13) and integrating, we obtain

$$n = \left(\frac{2\eta m}{3e^2\tau}\right)^{1/2} \frac{J^{3/2}}{s\dot{J}^{1/2}}. \tag{14}$$

It is assumed here that the rate of increase of the current is constant:

$$\dot{J} = \frac{U}{L} = \text{const},$$

where U is the initial voltage on the discharge capacitor. Using (13) and (14) we find for the rate of increase of the conductivity

$$\frac{d\sigma}{dt} = \left(\frac{3e^2\tau\eta}{2m}\right)^{1/2}\left(\frac{J\dot{J}}{s}\right)^{1/2}. \qquad (15)$$

Consequently, the thickness of the skin layer is determined as follows:

$$\delta = \frac{c}{\sqrt{4\pi}}\sqrt[4]{\frac{2m}{3e^2\tau\eta}}\left(\frac{J\dot{J}}{s^2}\right)^{-1/4} \qquad (16)$$

and therefore decreases with increasing current. Under our experimental conditions ($\dot{J} > 10^{10}$ A/sec) δ already becomes smaller than the radius of the discharge chamber for currents J = 100 to 200 A.

3. Constriction of the Plasma Sheath. As the discharge current increases, the plasma sheath, which comprises the skin layer, is constricted by the magnetic field associated with the current. The nature of the constriction is different, depending on the initial conditions. If it takes place in a time commensurate with the time $1/\nu$ of electron collisions resulting in ionization, the discharge can be constricted until complete ionization of the plasma sheath takes place. In slow constriction with a large rate of ionization the outer layers of the sheath can become ionized. The conductivity of the plasma no longer increases, and the skin effect disappears in these regions of the discharge. Then the discharge ostensibly grows toward the center, leaving the plasma with a rather high degree of ionization (> 5%) ("snowplow" effect), and the "ingrowth" rate is greater than the constriction rate. Then due to drifts the plasma can constrict into a slender filament. This is what is called the static pinch effect. Here we consider only the dynamic pinch effect realized in our own experiment.

The theory of the pinching of a plasma column was first developed by Leontovich and Osovets [53] and by Rosenbluth [58]. It was postulated in [53] that the discharge current increases linearly. This postulate is valid if the pinch time is considerably smaller than the half-period of the discharge current. Moreover, the constriction was assumed to be adiabatic. Under these conditions the constriction is described by the equation

$$\frac{d}{dt}\left(\frac{1}{3}M\frac{da}{dt}\right) = \frac{J^2}{ac^2} + 2\pi ap, \qquad (17)$$

where a is the radius of the plasma column, p is the initial pressure of the gas, and M is the mass of the gas per unit length of the plasma filament. We note that it is necessary to include on the right-hand side of (17), in general, the pressure increase due to heating of the gas during breakdown and during constriction. This consideration, however, does not introduce changes in the result if the magnetic pressure is many times larger than the gas-kinetic pressure. Assuming also that $\dot{J} = U/L$, we write Eq. (17) in the form

$$\frac{d}{dt}\left(M\frac{da}{dt}\right) = -\frac{3\dot{J}^2t^2}{ac^2} + 2\pi ap, \qquad (18)$$

where

$$M = \pi(a_0^2 - a^2)\rho_0.$$

Solving this equation simultaneously with the adiabatic equation

$$pv^{\gamma} = p_0v_0^{\gamma},$$

for a monatomic gas ($\gamma = 5/2$) we obtain the pinch time

$$t = \left(\frac{M}{3}\right)^{1/4}\left(\frac{ac}{j}\right)^{1/2}. \tag{19}$$

In the presence of excitation and ionization processes the quantity γ must differ from 5/2. Nevertheless, the pinch time calculated by this equation is usually consistent with the pinch times determined experimentally (see, e.g., [59]). In practical units the pinch time is equal to

$$t = 0.05 a_0 (p_0 \mu)^{1/2} j^{-1/2}. \tag{20}$$

The pinch time calculated by Rosenbluth with neglect of the pressure inside the unperturbed region differs only slightly from the results of the given calculation. It is important to note that these calculations were based on the assumption of complete entrainment of the neutral gas by the moving plasma sheath. As a rule, this assumption is valid, because the charge-exchange cross sections responsible for entrainment of the gas are large. For low densities, however, the mean-free path relative to the charge-exchange process can become comparable with the radius of the discharge chamber and the thickness of the plasma sheath. In this case the gas entrainment is incomplete, and the pinch time is smaller than the value given by Eqs. (19) and (20).

In the pinch effect there are large discontinuities of the temperatures and energies of directional motion of the electron and ion components of the plasma. In motion toward the center the ions acquire very large velocities, corresponding to energies of several tens and even hundreds of electron-volts. However, the thermal velocities of the ions are small. Thus, the heating of the ion component is caused primarily by energy transfer in collisions with electrons, which in turn are heated in the electric field. The equalization time of the electron and ion temperatures is large due to the large mass difference; it is determined as follows:

$$\tau = 17A\frac{T_e^{3/2}}{n}, \tag{21}$$

where T_e is the electron temperature and A is the atomic weight. For our experimental conditions we can roughly estimate this time. Taking $T_e = 3$ eV and $n = 3 \cdot 10^{14}$ cm^{-3}, we obtain $\tau = 10^{-5}$ sec, which is at least an order of magnitude greater than the pinch time calculated from Eqs. (19) and (20) or determined experimentally (see Chap. III). Consequently, the following relations hold during the pinch effect between the electron and ion temperatures and energies of radial motion:

$$E_i \gg kT_i, \quad kT_e, \quad T_e \gg T_i, \quad E_i \gg E_e,$$

where E_i and E_e are the energies of directional motion of ions and electrons, respectively. The directional energy of the electrons is considerably greater than that of the ions (in the ratio of their masses) or the thermal energy of the electrons themselves. The directional velocities of the electrons and ions during constriction remain identical, because any slight spatial separation of the components results in the production of electric fields which inhibit further separation. As a result, the plasma remains electrically neutral during constriction. The current density during constriction increases due to the increase in the total current and the decrease in diameter of the plasma sheath. Consequently, the temperature and degree of ionization of the plasma also increase. The variation of the electron temperature and certain other parameters, unfortunately, have only been calculated for the case of hydrogen [60, 61].

Another approach to the problem of the constriction of the plasma filament is also possible. Allen [62] has investigated the motion of a shock wave generated in the pinching of a plane current layer. Jukes [63] has analyzed the analogous problem for the case of cylindrical symmetry. The results of this analysis give a pinch time close to the time according to Eq. (19). Imshennik and D'yachenko [61] have considered the motion of a shock wave with cylindrical symmetry in a plasma, taking account of dissipative processes. It is important to realize, however, that it is not always justifiable to treat the constriction of the plasma sheath as shock-wave motion. Thus, a shock wave always represents a surface on which the main parameters of the medium vary abruptly in a layer of thickness equal to the mean-free path. Hence, it is only proper to speak of a shock wave when the mean-free path is smaller than the geometric dimensions of the medium. In our experiment, in particular, this requirement is not usually met.

4. Cumulative Discharge. As we showed above, there is a large discontinuity of the energies of the ion and electron components in the pinched plasma filament. This fact governs the thermodynamic equilibration dynamics at the instant of cumulative discharge. The main thing that happens is a rapid (in roughly the time of one collision) thermalization of the ion component. The temperature equalization of both components is rather slow. It requires $\sqrt{M/m}$ collisions. In the adiabatic approximation the temperature equalization time is determined by the following equation, which holds for any ratio of the electron and ion velocities:

$$\tau = \frac{3}{8\sqrt{2\pi}} Mm\sqrt{k^3}\frac{1}{ne^4\Lambda}\left(\frac{T_e}{m}+\frac{T_i}{M}\right)^{3/2}, \tag{22}$$

in which M is the ion mass and k is the Boltzmann constant. The time τ is now considerably smaller than in pinching, due to the increase in the electron density. In the temperature equalization process the electron temperature and degree of ionization increase rapidly. Although the ion energy is considerably greater than the ionization potential during pinching, the ionization cross section associated with ion collisions is more than an order of magnitude smaller than the electron-impact ionization cross section. Moreover, the ion velocities are smaller than the electron velocities by at least two orders of magnitude. Consequently, the role of ion collisions in the excitation and ionization processes can be neglected. The plasma temperature at the instant of cumulative discharge is several times the equilibrium temperature at the current densities that exist in the plasma filament (see [59] for the hydrogen case). We also point out that throughout the entire temperature equalization process there is a large temperature differential between the electron and ion components on account of the electron losses in ionization and excitation of multielectron atoms. As a result, the temperature equalization time is increased (see more in detail in [64, 65]). The large temperatures and degree of ionization in the pinched filament are also evinced by the experimentally observed bremsstrahlung burst and spectral lines of multiply charged ions (see, e.g., [66]).

5. Disruption of the Plasma Filament with the Growth of Instabilities. Following cumulative discharge the plasma filament again begins to expand. However, the electrodynamic forces now prevent expansion, and for a certain time the constriction process is again initiated. After two or three constrictions the filament disintegrates due to the development of magnetohydrodynamic instabilities. Kinking of the plasma filament is observed even in the first pinch. Mainly necking-down instabilities occur at this time ("sausage" instability). In discharge accumulation, in the case of a low density, countercurrent beams occur, which lead to beam instability. Then other types of instability set in ("serpentine" or "corkscrew" instability, convective instability, etc.).

Let us assume that the plasma filament has a stable configuration. Suppose also that the surface of the filament suffers a certain perturbation, which we represent in the form

$$q = q_0 e^{i(m\theta+kz)+\omega t},$$

where q is the filament radius or the local electric or magnetic field, m = 0, 1, 2, 3, ... for different types of instabilities, and k is a wave number characterizing the longitudinal periodicity of the perturbation. The quantity $e^{\omega t}$ describes the growth rate of the instability. For the analysis of instabilities in the set of equations describing the state of the plasma (Maxwell equations, continuity equations, Euler equation, and boundary conditions) it is required to substitute the perturbation values of E, H, and j in the form

$$H = H_0 + H_1 e^{i(m\theta + kz) + \omega t}$$

(E and j are written analogously). The values of m and k giving positive real values of ω express unstable perturbations.

In a pinch discharge without an external magnetic field long-wave perturbations dominate, for which m = 0. This instability is called a "sausage" instability. It results in the formation of necks in the filament and a large potential drop lengthwise. There occur as a consequence large local discontinuities in the electric and magnetic fields. The acceleration processes attending this effect give rise to a burst of neutron radiation, which has been observed experimentally [67]. Perturbations with m = 1 produce bends in the filament and dominate in the presence of an external magnetic field. For lack of space we shall not discuss other types of instability.

§2. Possible Mechanisms of the Formation of Negative-Temperature States in the Pinch-Discharge Plasma

It is possible to conceive several ways in which population inversion can develop in a pinch-discharge plasma. In general, population inversion can occur even in the first stage of discharge, as in a number of pulsed lasers. For small initial densities, as in our experiment, however, the gain can prove inadequate to overcome the losses in the cavity.

During pinching of the plasma sheath the plasma density and temperature increase, and at a certain stage the gain becomes sufficient for the onset of lasing. The required excitation intensity depends, of course, on the specific quantum system (we give a detailed discussion for the case of argon at the end of the present article). In the presence of a shock wave lasing can be initiated at the shock front due to rapid heating. One mechanism for the stimulation of emission, clearly, could be the Penning effect, when discharge is realized in a mixture of suitably chosen gases. The latter can be, for example, mixtures of helium and xenon or of neon and krypton, in which the Penning discharge takes place.

The most auspicious from the standpoint of creating negative-temperature states appear to be the processes attending cumulative discharge. The large discontinuity of the ion and electron temperatures and various instabilities create extreme deviations from thermodynamic equilibrium. It is also important to mention the large plasma density, which promotes considerable population inversion, although at the same time it must lead to fast relaxation. Usually at the time of cumulative discharge there is a prevalence of ions having different ionization multiplicities. It is in the system of levels of such ions that population inversion takes place. Consequently, lasing should be expected predominantly in the ultraviolet region of the spectrum.

§3. On the Experimental Program

As we showed above, the pinch-discharge plasma comprises a medium with strongly pronounced deviations from thermodynamic equilibrium. However, it is not obvious *a priori* that the deviations from equilibrium are large enough to create population inversion between definite energy levels. It is even less likely that population inversion states will exist with sufficient frequency. Naturally the presence of such states occurs only at definite times and in a definite region of space and must depend on the initial conditions and specific quantum system. Due to the high temperature and degree of ionization of the pinch-discharge plasma it seems logical to assume that negative-temperature states are most probable in the level systems of singly and multiply ionized atoms. In the early part of our experiments lasing was realized only from mercury ions. We therefore sought to stimulate emission in various gases, primarily the inert gases. The experiments were conducted in large-diameter (100 mm) chambers, where they proved fruitless, mainly due to the fact, as will be demonstrated presently, that the interval of optimum pressures for lasing in large-diameter (>40 mm) tubes is outside the breakdown voltage range. Our subsequent attempts to effect lasing were carried out in small-diameter (10 mm) tubes [31] with argon ions, at whose transitions we were able to realize lasing in ordinary pulsed discharges [9-11]. In small-diameter tubes the discharge had an intermediate behavior between ordinary and pinch discharges. Then, in order to isolate the effect in pure form, we investigated lasing in large-diameter tubes under conditions such that the pinch effect had a distinct character. The experimental apparatus was constructed so that the discharge conditions could be varied over a wide range. The nature of the discharge was monitored by various measurements, including probe measurements of the current-time distribution, high-speed photoscanning of the discharge cross sections, and plasma diagnostics based on spectral methods. This kind of investigation was mandatory, because the voluminous literature on the subject refers primarily to hydrogen pinch. The data on argon discharges, on the other hand, generally refer to conditions altogether different from those realized in our experiment.

CHAPTER III

INVESTIGATION OF THE CHARACTERISTICS OF A PINCH DISCHARGE

§1. Description of the Experimental Setup

A schematic diagram of the experimental arrangement used to investigate stimulated emission in a pinch discharge is shown in Fig. 4. The energy source is the capacitor C, which is charged by the high-voltage rectifier HVR. Current switching is realized by the spark gap G, and the actual discharge takes place in the discharge chamber DC between electrodes E.

We used 0.01- to 2.5-μF capacitors with voltages from 15 to 45 kV. The upper voltage limit was determined by the capabilities of the HRV, and the lower limit by the breakdown conditions. The fundamental principles governing the construction of the experimental apparatus are operational reliability and the ability to make quick changes in the experimental conditions. These requirements are typical of exploratory experiments. We fabricated three discharge chambers, which were mounted on a single stand. The construction of the first chamber en-

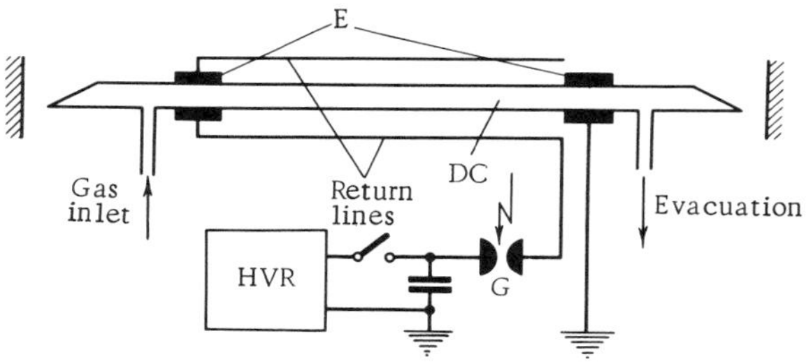

Fig. 4. Schematic of experimental arrangement.

abled us to rapidly interchange tubes with diameters ranging from a few millimeters to 2 cm. The diameter of the second chamber was 25 to 30 mm, and that of the third was 40 mm. The length of all the chambers was 100 cm. The working-gas evacuation, intake, and power supply systems were common to all the chambers. The changeover from one chamber to another was realized by switching of the cable conductors.

Unlike standard discharge chamber designs, the electrodes were made in the form of rings with an inside diameter equal to or greater than the diameter of the discharge tube. This feature allows for the unobstructed efflux of radiation in the axial direction while maintaining the axial symmetry of the discharge. For loss reduction the exit windows were oriented at the Brewster angle; they were made of glass or quartz. To preclude contamination and damage during discharge the windows were spaced 15 to 20 cm from the electrodes. To reduce contamination of the discharge by vaporization of the material of the electrodes the latter were usually coated on the inside with tantalum or molybdenum. In all the experiments we used quartz discharge tubes. The tube was vacuum-coupled with the electrodes either by means of special teflon seals or with epoxy resin. To decrease the inductance of the discharge chamber a coaxial system of return leads was used, comprising four to six copper rods situated in the immediate vicinity of the tube surface. Cables or busses were used in the other elements of the discharge circuit. Switching of the discharge current was effected by means of low-inductance vacuum spark gaps, which enabled us to work with currents in the hundreds of kiloamperes. The chamber was furnished with the capability for preionization of the working gas. A UVCh-4 oscillator operating at a frequency up to 30 MHz and power of 90 W was used for this purpose.

The gas evacuation and intake systems were made of glass and joined to the discharge chamber at opposite ends. One-shot gas injection was normally used. We used only spectrally pure gases. The chamber was evacuated with a fore pump and diffusion pump using a nitrogen trap, creating a vacuum of $\sim 10^{-6}$ torr.

For the optical cavity we used either plane mirrors or a nearly confocal system of mirrors. In the latter case the radius of curvature of the mirrors was 3 m, and their mutual separation was 140 to 150 cm; the mirrors were 40 to 50 mm in diameter. We normally used mirrors with dielectric coatings. We had at our disposal a supply of mirrors with reflectivities ranging from 30 to 99% in the interval of wavelengths from 3400 to 6000 Å. The mirrors were mounted in special holders, which could be moved perpendicularly to the optic axis of the cavity. Additional adjustments of the mirrors were practically unnecessary after their displacement.

A shortcoming of the system described above is the impossibility of varying the discharge current growth rate over very wide limits, and this variable is one of the basic quanti-

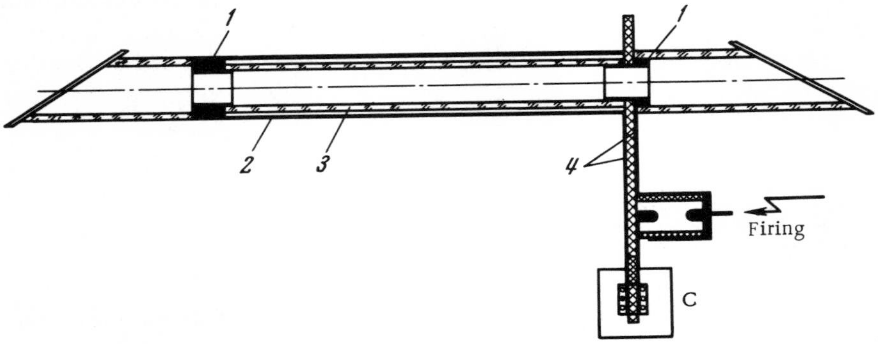

Fig. 5. Schematic of apparatus with low stray inductance.
1) Electrodes; 2) metal tube; 3) quartz tube; 4) bus.

ties governing the pinch dynamics. The current growth rate is $\dot{J} = U/L$, where U is the initial voltage and L is the discharge circuit inductance. Thus, for a given L the interval of variation $\dot{J}$ is determined by the initial voltage. We were able, however, to increase the current growth rate substantially by decreasing the stray inductance of the discharge circuit, as this quantity was several times the inductance of the discharge chamber. For this purpose we built an apparatus (Fig. 5) having a very small stray inductance (considerably less than the discharge inductance), thus effecting a four- to sixfold reduction in the total inductance of the discharge circuit and a proportionate increase in the current growth rate. For the return conductor we used copper tubing, into which was inserted a quartz discharge tube 25 mm in diameter. The end of the tube served as one of the electrodes. A coaxial air gap was used for current switching. The power supply was one or two parallel-connected low-inductance 0.1-μF capacitors. Copper busses were used for the leads. Only a relatively small fraction of the results were obtained on this apparatus. All of the experiments, unless specifically stated otherwise, were conducted on the apparatus described at the beginning of this section.

§2. Discharge Electrotechnical Parameters

A Rowgowski loop was used to record the discharge current. The signal from the loop was sent through an integrating network to an oscilloscope. The sensitivity of the loop is equal to

$$\frac{U}{J} = \frac{R_I}{N} k,$$

where U is the signal from the loop, N is the total number of turns, k is the number of turns of the loop around the current-carrying conductor, and R_I is the integrating resistance. We used a loop with the following parameters: N = 830; $R_I = 2\Omega$; k = 5; J/U = 83 *a/b*. A special shielding system was used to reduce the stray inductance without diminishing the sensitivity of the loop.

A typical current oscillogram is shown in Fig. 6. The current pulse is sinusoidal with large damping. Four or five half-periods are usually witnessed. The period and maximum current values for different capacitances at an initial voltage of 25 kV are given in Table 2.

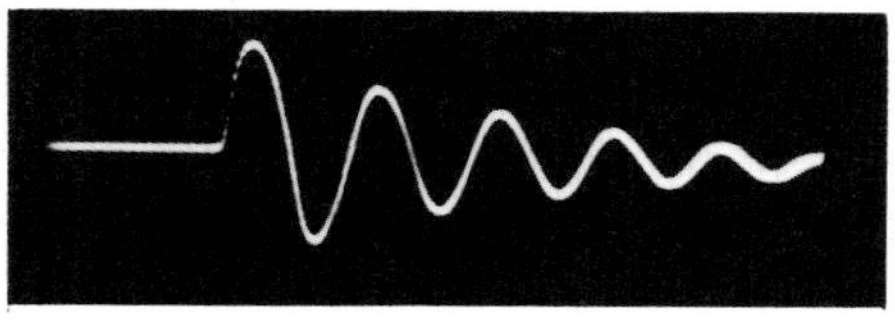

Fig. 6. Typical discharge-current oscillogram.

TABLE 2

Capacitance, μF	Period, μsec	Current, kA	Period, μsec	Current, kA
0.01	0.9	1.1		
0.1	2.5	4.5	1.0	8
0.2	3.5	5.5	1.4	10
0.4	5.0	7.5		
2.5	12.5	13		

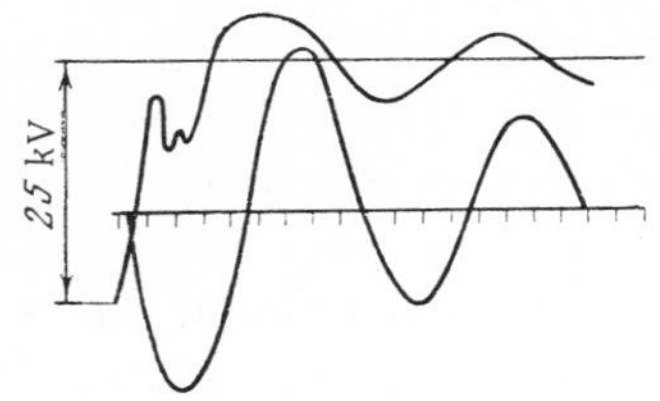

Fig. 7. Oscillogram of voltage pulse across discharge chamber (upper curve) and current (lower curve) in a 20-mm-bore tube ($C = 0.1\mu F$; $p = 10^{-2}$ torr).

The reduced values are approximately identical for tubes of different diameters and at different initial pressures. The values in the fourth and fifth columns refer to the low-stray-inductance system.

To measure the voltage drop across the discharge chamber we used a resistance voltage divider connected in parallel with the discharge gap. The resistance of the lower arm of the divider was determined by the matching conditions with the recording system. All other requirements imposed on measurements of this type were satisfied in this case. In addition to the voltage we recorded the discharge current. The current and voltage curves (Fig. 7) coupled with data on the discharge geometry according to high-speed photoscanning of the discharge make it possible to determine such parameters of the discharge as its conductivity and electron temperature and the electrical energy input to the discharge. These measurements will be described in detail in the fifth chapter.

§3. Probe Measurements of Current Distribution over the Discharge Tube Cross Section

The pinch effect is characterized by a rapid variation of the cross-sectional distribution of the current in the discharge tube. Consequently, the current distribution measured experimentally gives direct information about the nature of the discharge. A miniature Rogowski loop that could be inserted into the discharge and displaced over its cross section was specially fabricated for the measurements (Fig. 8). This loop permits a direct determination of the current density at different points of the discharge at different times. The probe construction is distinguished by a high signal-to-noise ratio, which is achieved mainly by double-row winding of the loop. A description and detailed calculation of the loop are given in [68].

In order to realize sufficient spatial resolution we conducted the experiments in a large-bore (40 mm) chamber. This feature made possible five or six nonoverlapping measurements over the chamber cross section. Provision was also made for variation of the probe orientation relative to the current direction. The latter capability is necessary in order to monitor the noise intensity; when the probe is aligned with the current, the discharge current is not recorded and the probe signal can be used to estimate the magnitude and character of the stray inductances. The probe pulse was recorded by means of an OK-17M double-beam oscilloscope, to the input of which was transmitted the signal from the Rogowski loop used to record the total current in the discharge circuit. In several cases an S-I-29 oscilloscope with "memory" was

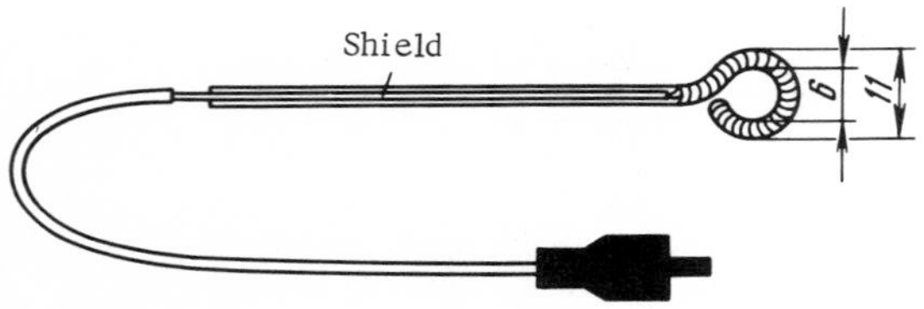

Fig. 8. Miniature Rogowski loop for measurement of cross-sectional current distributions in the discharge chamber.

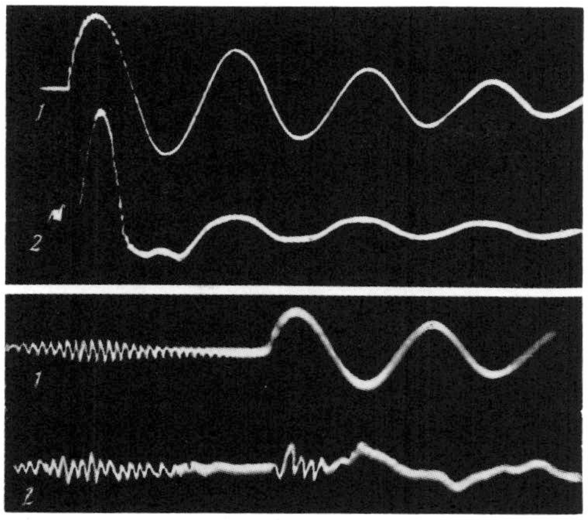

Fig. 9. Oscillograms of probe curves (2) for two probe positions: center of the chamber (upper curve) and at the wall (lower curve) under near-optimum lasing conditions (C = 0.4 μF; U = 25 kV; bore = 40 mm; p = 10^{-3} torr). Curves 1 represent the total current.

used. Typical oscillograms obtained with the probe in two extremal positions – in the center of the chamber and at the wall – are shown in Fig. 9. The oscillograms were obtained under essentially optimum lasing conditions: U = 25 kV and p = 10^{-3} torr. Characteristic jumps corresponding to successive pinches of the discharge are visible in each half-period of the oscillogram for the central position of the probe in the discharge. In the first pinch practically the entire current is concentrated in the center. Thus, near the wall the probe does not put out a signal. In subsequent half-periods the current jumps in the center of the chamber are less pronounced, while the current at the wall increases to the value recorded by the probe. In each half-period there is only one pinch, a result that is attributable to the short current rise time. The duration of the constriction phase is 0.4 μsec.

The nature of the probe curves changes with the initial discharge conditions, i.e., with the pressure and voltage. Oscillograms of probe pulses recorded at various initial pressures are shown in Fig. 10. The clearest pattern corresponding to one pinch occurs at a pressure $3 \cdot 10^{-3}$ torr, which is the optimum for lasing. With an increase in pressure the nature of the curves is altered: The envelope of the current pulse has superimposed on it a high-frequency background,

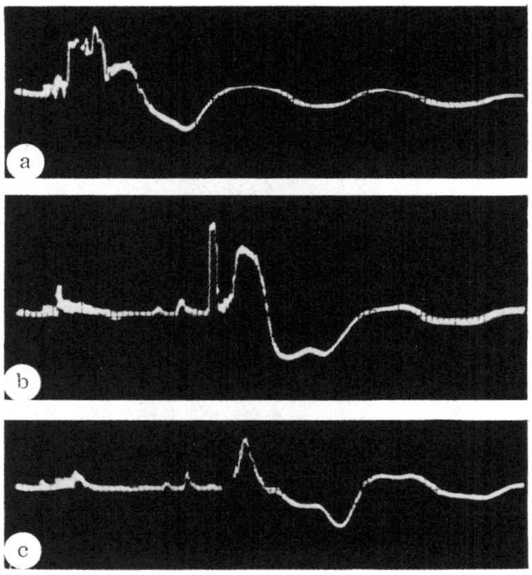

Fig. 10. Oscillogram of current at the center of the discharge chamber for various initial pressures (C = 0.4 μF; bore = 40 mm; U = 25 kV). a) 10^{-1} torr; b) 10^{-2} torr; c) $3 \cdot 10^{-3}$ torr.

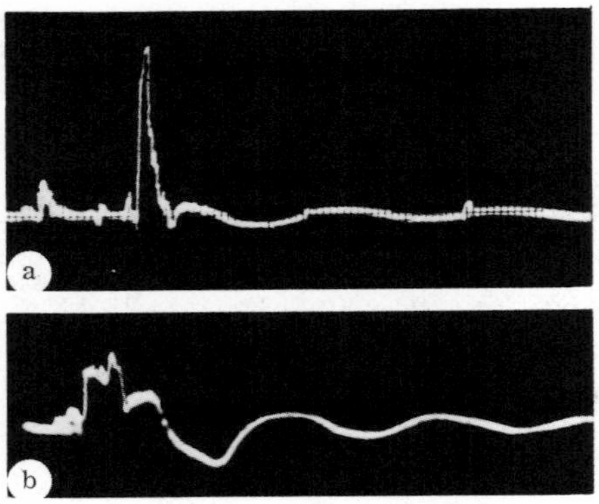

Fig. 11. Oscillogram of current at the center of the discharge chamber for various initial voltages ($p = 10^{-1}$ torr; $C = 0.4\,\mu F$; bore = 40 mm). a) U = 10 kV; b) U = 25 kV.

which possibly corresponds to instabilities setting in during the pinch effect itself. For low voltages the background intensity decreases, and the probe curves have the same form as for low pressures (Fig. 11).

§4. High-Speed Photoscanning of Pinch Radiation

The luminescence of the discharge was recorded with a high-speed photoscanning (HSPhS) camera having a time resolution of $\sim 0.1\ \mu$sec in the single-exposure mode. The sensitivity of the camera was increased by means of special optical intensification attachments, which permitted the discharge to be photographed at close range (0.5 m). The photographs were made in both the single-exposure and frame-by-frame scanning modes, in either case at the side of the discharge chamber (Fig. 12). High-speed photoscanning was performed for the chambers of various diameters used in the investigation. In all the chambers of diameter greater than 10 mm the pinch effect was clearly pronounced when a capacitor of $\geq 0.1\ \mu F$ capacitance was used. With a reduction in capacitance to 0.01 μF the current rise time is too small, and complete pinch does not take place. In this case the nature of the discharge is similar to that described in [69]; the discharge is deflected from the walls and subsequently bends. The pinch times measured from the photoscans for different chambers in the interval of optimum pressures for las-

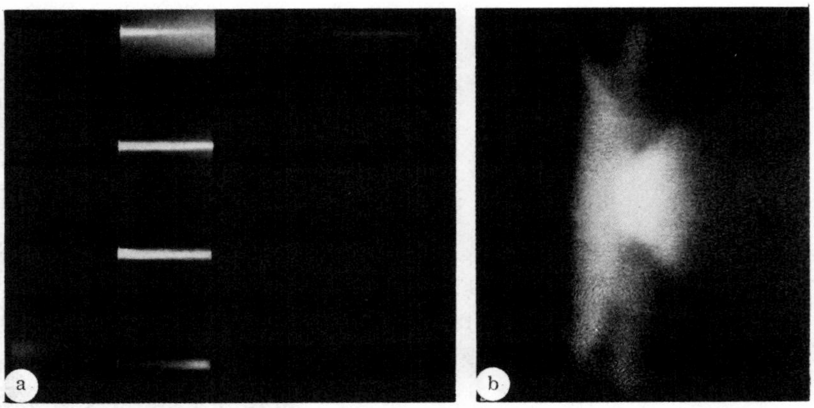

Fig. 12. Typical HSPhS camera exposures of pinch radiation recorded in frame-by-frame scanning (a) and in the single-exposure mode (b) ($C = 0.4\,\mu F$; U = 25 kV; $p = 10^{-2}$ torr; bore = 30 mm).

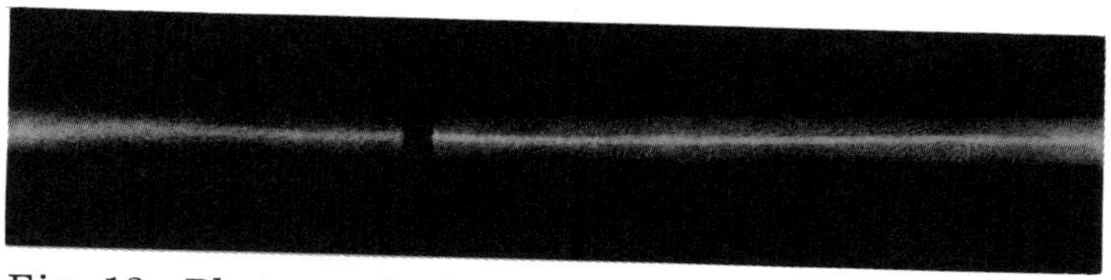

Fig. 13. Photograph of pinch in a 20-mm-bore chamber without time scanning (C = 0.1 μF; U = 25 kV; p = 10^{-2} torr).

ing are somewhat shorter than the times calculated according to Eq. (20) of Chapter II. This result is caused by the incomplete entrainment of the neutral gas by the constricted plasma sheath at the relatively low pressures corresponding to the lasing interval. The diameter of the plasma filament at the instant of cumulative discharge is 2 to 4 mm, and the discharge persists in the pinched state for 0.3 to 0.4 μsec according to the probe measurements. Pinches corresponding to subsequent current half-periods are also visible in the HSPhS exposures, but the luminous intensity in this case is small, so that detection is difficult. We note that the luminous intensity from the central region of the discharge is many times the luminous intensity of other parts of the discharge. In the photographs recorded without time scanning (in the first experiments), therefore, a distinct filamentation of the discharge is visible. The filament diameter in this case is close to the discharge diameter at the instant of the cumulative effect as determined from the HSPhS exposures (Fig. 13).

§5. Discharge Spectral Characteristics

As we know, in the pinch discharge a considerable fraction of the input energy is lost in radiation in the visible and ultraviolet regions of the spectrum. The pinch spectrum comprises a line spectrum against a continuous-radiation background. The continuous spectrum occurs at the instant of cumulative discharge and is caused chiefly by bremsstrahlung. Ordinarily the pinch spectrum contains lines associated with impurities, mainly silicon, that emanate from the chamber walls. As a rule, these lines appear after the first pinch.

In the present investigation we recorded the time-integral spectrum. The individual lines were scanned by photoelectric techniques. An STÉ-1 spectrograph, which has a relatively good dispersion of 5 to 10 Å/mm in the wavelength interval of 2000 to 6000 Å, was used to record the spectrum. The nature of the spectrum recorded at the side and ends of the discharge chamber is the same, and signs of resorption were not in evidence. Identification of the spectrograms indicated the following:

1. The most intense lines belong to argon ions with different ionization multiplicities: Ar II and Ar III.
2. The intensity of even the strongest lines for neutral argon is insignificant.
3. Exceedingly weak lines are present due to impurities, chiefly nitrogen, oxygen, and silicon ions.
4. As a rule, weak lines are observed for hydrogen H_α and H_β.

Inasmuch as the current in the wall region is very slight during the first half-period, as indicated by the probe measurements, wall impurities do not enter into the discharge during this time. Consequently, the impurities do not affect the character of the emission (observed in the first half-period).

A system of diaphragms 2 mm in diameter was used for time scanning of the emission of individual lines at different points of the discharge across the diameter of the chamber (see

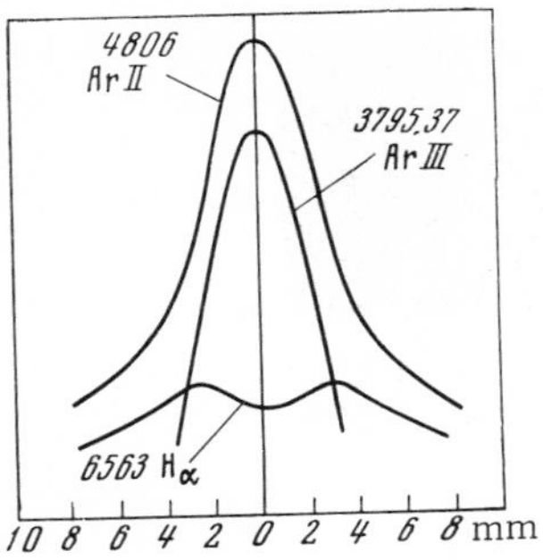

Fig. 14. Intensity distribution of various lines over the discharge diameter.

Fig. 28, Chapter V). The individual lines were resolved by means of a ZMR-3 monochromator with a photomultiplier at its output. The intensity distribution for the individual Ar II and Ar III lines over the diameter of the 2-mm chamber is shown in Fig. 14. The intensity distribution of the Ar III lines refer to the instant of cumulative discharge. The intensity of the Ar II line refers to the constricted plasma sheath, so that the various points of the curve correspond to different stages of pinch. Also shown in the figure, along with the argon lines, is the intensity distribution for the hydrogen H_α line. The intensity of this line at the center of the chamber, corresponding to cumulative discharge, is small, despite the fact that the plasma density in this period of time greatly exceeds the initial value. The low intensity of the hydrogen lines indicates a high temperature and large degree of ionization at cumulative discharge.

CHAPTER IV

PROPERTIES OF STIMULATED EMISSION IN A PINCH-DISCHARGE PLASMA AT THE TRANSITIONS OF SINGLY AND DOUBLY IONIZED ARGON ATOMS

The first results bearing on lasing in a pinch discharge were obtained in a tube 10 mm in diameter [31]. We showed in the preceding chapter that a clearly pronounced pinch is already observed in tubes of this diameter. For the further elucidation of the role of the pinch effect lasing was investigated in both large- and small-bore tubes. As a result, various characteristics (temporal, spatial, spectral, etc.) of the emission process were obtained as a function of the parameters of the pinch: the pressure of the working gas, voltage, capacitance of the capacitor, discharge current growth rate, and bore diameter of the discharge chamber. The quantity most difficult to vary is the current growth rate

$$\dot{j} = U/L,$$

where U is the initial voltage and L is the proper inductance of the discharge circuit. The possibility of varying $\dot{J}$ by changing the voltage is extremely limited. Also, this technique does not guarantee reproducibility of the initial conditions, so that, for example, the conditions of the first phase of discharge and, hence, of the ensuing stages can change. In order to explain the role of $\dot{J}$, therefore, it is necessary to change only the inductance of the discharge circuit. The two discharge circuit designs described in the preceding chapter have inductances that differ by a factor of four to six. The values of $\dot{J}$ differ by the same amount. The bulk of the measurements in both cases refer to an initial voltage of 25 kV.

§1. Properties of Emission by Ar II Ions

Under pinch conditions the greatest intensity always occurs at the 4764.89 Å line. Consequently, all measurements described below refer to precisely this line. In this section we give certain characteristics of the lasing process as a function of the discharge conditions.

a. Temporal Characteristics of the Discharge. The principal objective in the first experimental phase was to establish a time correlation between the instants of lasing and pinch. For this purpose we recorded the emission pulse simultaneously with the current so as to be able to determine the delay time of lasing with respect to the initiation of discharge. We then compared these times with the data obtained by high-speed photoscanning (HSPhS). The emission pulse was recorded with a photomultiplier and delivered to the input of an OK-17M oscilloscope. The signal from the Rogowski loop was sent to the second input of the oscilloscope. In a number of cases an FÉK-12 photocell was used to record the emission, so that the signal could be transmitted directly to the oscillograph deflection plates. In this case we usually used an OK-21 oscilloscope, which has a strong beam brightness for short sweeps. A typical oscillogram of a current and emission pulse is shown in Fig. 15. Lasing is initiated with a certain delay, different for different tube diameters, relative to the discharge current. The delay time is independent of the capacitance for $C \geq 0.1\ \mu F$ up to $2.5\ \mu F$. This independence is entirely reasonable, because the behavior of the current curves up to the inception of lasing is approximately the same for different capacitances. The delay times for tubes of various diameters are given in Table 3. Also shown in the table for comparison are the pinch times determined from the HSPhS exposures.

Thus, it is evident from the table that lasing is initiated just before the time of cumulative discharge. This conclusion is further evinced by the fact that the diameter of the emission spot greatly exceeds the diameter of the plasma filament at the time of cumulative discharge. The diameter of the emission spot was determined with an ordinary camera. It coincides with the active medium diameter, which is determined more precisely by the insertion of a miniature diaphragm into the plane cavity, where it could be moved perpendicularly to the optic axis of the cavity.

With variation of the initial voltage of the lasing initiation time also changes. Thus, for a tube 30 mm in diameter this time decreases from 0.35 to 0.25 μsec as the voltage is increased from 15 to 35 kV. The pinch time varies in approximately the same ratio. This cor-

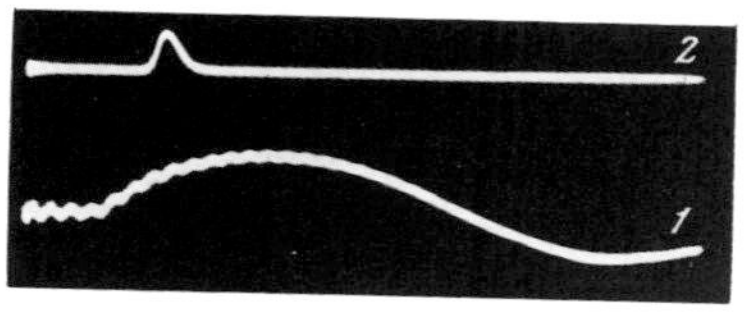

Fig. 15. Simultaneous oscillogram of discharge current (1) and emission pulse (2) in a 30-mm-bore chamber ($C = 0.1\ \mu F$).

TABLE 3

Discharge chamber diameter, mm	Lasing delay time relative to current, μsec	Pinch time, μsec	Emission spot diameter, mm
25	0.25—0.30	0.35—0.40	12
30	0.3—0.35	0.40—0.45	15
40	0.4	0.50—0.55	25
10	0.1	—	10
15	0.1—0.15	0.20	10
20	0.2—0.25	0.30—0.35	12

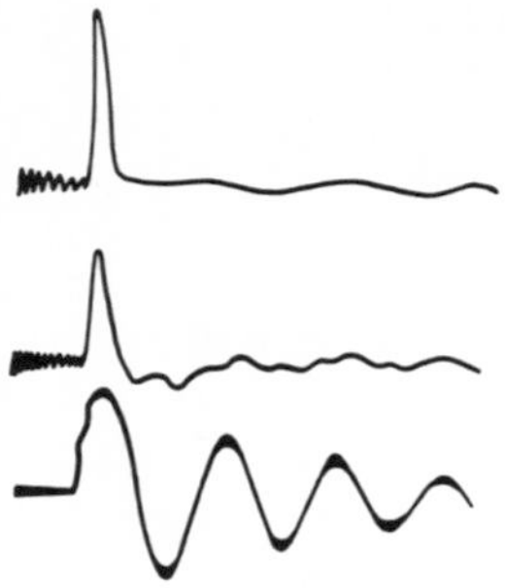

Fig. 16. Oscillogram of current (a), probe pulse from the center of the discharge chamber (b), and laser emission (c) (bore = 40 mm; C = 0.1 μF; U = 25 kV; p = $2 \cdot 10^{-3}$ torr.

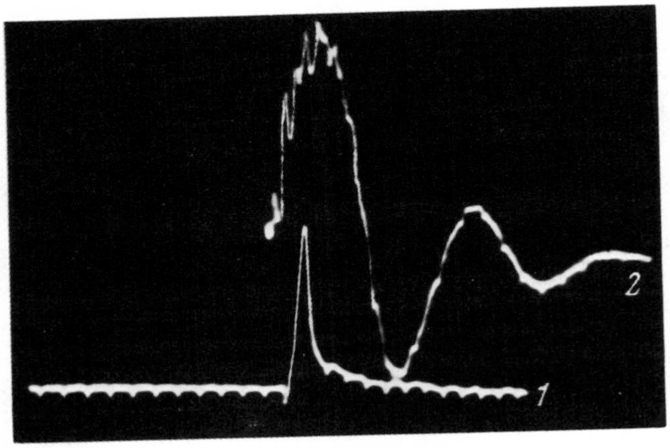

Fig. 17. Oscillogram of emission pulse (1) and current (2) in a 25-mm-bore tube on the low-stray-inductance apparatus (C = 0.2 μF; U = 25 kV; p = $8 \cdot 10^{-3}$ torr; sweep time, 4 μsec.

respondence is best exhibited by a comparison of the data obtained for systems with different current growth rates. With a fourfold increase in $\dot{J}$ the pinch time is cut in half according to Eq. (20) of Chap. II. The delay time of lasing relative to the current start time is also roughly halved; it is equal to or slightly less than 0.2 μsec for a 25-mm-bore tube.

The correlation between lasing and pinch is also exposed by probe measurements. The appropriate oscillograms are shown in Fig. 16.

The width of the emission pulse is almost constant for tubes of all diameters, being equal to 0.2 μsec at the half-peak intensity level. The intensity decreases at the time of cumulative discharge, but it does not vanish altogether until after 0.6 μsec. For large-bore tubes the emission pulse width is also independent of the initial pressure. For small-bore tubes a dependence is observed. With an increase in $\dot{J}$ the emission period at the half-peak level decreases. Thus, with a fourfold increase in $\dot{J}$ the pulse width decreases to 0.13 μsec for a 25-mm-bore tube. In this case as well, however, lasing does not stop until after 0.5 or 0.6 μsec (Fig. 17).

b. Initial-Pressure Dependence of Emission Intensity for Chambers of Various Diameters. The emission intensity was measured as a function of the initial pressure for tubes of diameter 7, 8, 10, 15, 20, 25, 30, and 40 mm. The measurements were performed with a photomultiplier preceded by neutral light filters for attenuation of the signal to the level corresponding to the linear interval of the photomultiplier characteristic. One emission line (4764.89 Å) was also separated by light filters. All the measurements were performed with a single photomultiplier, whose sensitivity was periodically checked. The initial conditions, i.e., the voltage and capacitance (25 kV and 0.1 μF), as well as the parameters of the optical cavity were identical in all the experiments. The cavity was formed by two dielectric mirrors, one plane and one spherical, with reflectivities of 99 and 80%, respectively, at the emission wavelength. These cavity parameters were optimum for all the tubes, since the gain is virtually independent of the chamber diameter. Curves of the emission intensity for tubes of various diameters are shown in Fig. 18. As the curves in this figure reveal, with increasing chamber diameter the interval of optimum values is shifted toward lower pressures and becomes narrower. In going from a 10-mm- to a 40-mm-bore tube the width of the optimum pressure interval and the optimum pressures themselves are decreased by about an order of magnitude. The emission intensity, on the other hand, increases sharply with the diameter. The maximum emission intensities for the 4764.89 Å line (in relative units) and their corresponding pressures are summarized in Table 4.

The emission intensity increases approximately as the square of the chamber diameter (and as the diameter of the emission spot). The decrease in intensity in the 40-mm-bore tube is clearly due to contamination of the discharge due to the low initial pressures and poor breakdown. Consequently, the emission intensity in a pinch discharge in a 30-mm-bore tube is one and a half orders of magnitude greater than in a 7-mm-bore tube, in which the conditions are close to those of conventional pulsed lasers. The peak radiated power in our experiment, measured with a calibrated photocell, is approximately 1 kW.

With an increase in $\dot{J}$ the optimum pressure interval is almost unchanged, although the emission intensity does increase (1.5-fold when $\dot{J}$ is quadrupled).

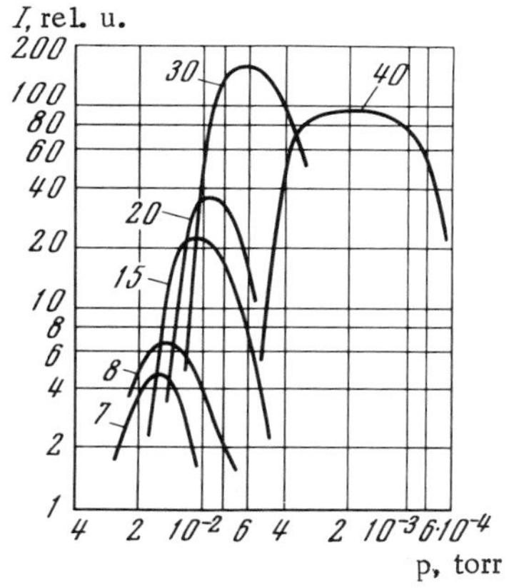

Fig. 18. Emission intensity versus initial pressure in chambers of various bore diameters (C = 0.1 μF; U = 25 kV).

TABLE 4

Discharge chamber diameter, mm	Emission intensity, relative units	Optimum pressure, 10^{-3} torr	Discharge chamber diameter, mm	Emission intensity, relative units	Optimum pressure, 10^{-3} torr
7	50	20	20	360	10
8	70	20	30	1400	6
10	150	15	40	900	2
15	260	12			

c. Spectral Composition of the Ar II Stimulated Emission in a Pinch Discharge. As mentioned in Chapter I, the intensity ratio of the 4764.89 and 4879.9 Å lines is fundamental to the explication of the lasing mechanism in argon. Dominant emission at the first line is related to direct excitation, and at the second line to step-by-step excitation. In our experiment the intensity ratio of these lines was investigated in tubes of different diameters under different initial conditions. In order to preclude the possibility of these line intensities being affected by emission at other transitions having a common upper or lower level, we decided upon a cavity with maximum reflectivity in a relatively narrow wavelength interval, from 4700 to 4900 Å. The reflectivities for both wavelengths were identical for each mirror and equal to 99 and 83%. The individual lines were recorded either with a monochromator and photomultiplier or by means of a spectrograph. The measurements showed that the intensity of the 4879.9 Å line is always one or two orders of magnitude lower than for the 4764.89 Å line or is in fact equal to zero.

Special attempts to realize emission at other transitions were not included in our problem. However, limited studies in the interval from 4200 to 5200 Å with mirrors 90 to 99% reflecting in this interval proved fruitless. Not to be dismissed, of course, is the possibility of emission at other transitions with a more careful choice of discharge conditions.

d. Certain Properties of Emission at the 4764.89 Å Line. Gain. The approximate value of the gain can be determined from the value of the cavity losses at which emission ceases. The reflectivity of the mirrors must be close to 100%. We used this procedure to deduce that the gain at pressures corresponding to maximum emission intensity is equal to 10 dB/m and depends only slightly on the chamber diameter.

Divergence and Structure of the Emission Spot. The divergence is maximal for large-bore tubes. With the use of a confocal cavity it attains 1°. For a semiconfocal mirror system the divergence is several times smaller.

The structure of the emission spot depends on the discharge conditions, mainly the initial pressure. This dependence is the most pronounced for small-bore tubes, with their wide range of optimum pressures. In the center of the optimum pressure interval the emission spot is circular and has uniform brightness over its entire cross section. With an increase in pressure the brightness of the spot becomes nonuniform, being a maximum at the center and at the periphery. For still higher pressures the emission has the shape of a thin ring or a tiny spot in the center. At below-optimum pressures the emission has the shape of a small spot diffuse at the edges (Fig. 19). A nonuniform brightness distribution over the cross section of the emission spot has also been observed in [70]. The emission normally had a ring configuration. The absence of emission inside the ring was attributed to resonance capture of the radiation, resulting in overfilling of the lower laser level. The more intricate structure of the emission spot in our experiment is caused by the nonuniform distribution of the plasma density over the chamber cross section. Resonance trapping of the radiation can occur in this case as well.

Influence of Preionization. Preionization, which alters the initial conductivity of the medium and breakdown conditions, can in general change the whole character of the evolution of discharge. In our experiments we realized preionization by means of a very low-power UHF oscillator, so that it clearly did not have an appreciable effect on the nature of the discharge. It was noted, however, that in small-bore tubes with preionization the emission intensity drops somewhat. Moreover, in the presence of preionization the reproducibility is always improved, the emission intensity remaining virtually constant in the course of many "shots" at a repetition rate of several hertz. Without preionization the emission intensity rapidly decreases. This effect can be attributed either to contamination of the discharge or to a decrease in the gas density on account of absorption by the chamber walls (hardening). In the presence of hardening the role of preionization is tantamount, clearly, to degassing of the chamber walls

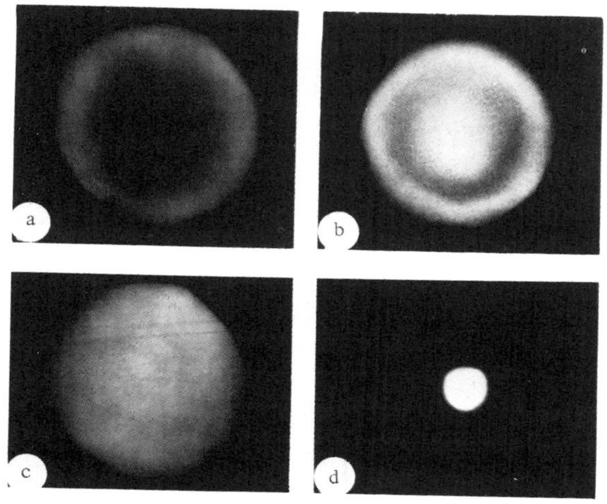

Fig. 19. Structure of the emission spot in a tube 20 mm in diameter at various initial pressures. a) p = 1.7 · 10^{-2} torr; b) p = 1.3 · 10^{-2} torr; c) p = 1 · 10^{-2} torr; d) p = 6 · 10^{-3} torr.

and restoration of the initial density of the working gas. Gas absorption at the walls can also be explained by the fact that at pressures above the optimum values there is no emission pulse in the first "shot," but there are pulses in the subsequent shots. The most probable mechanism of this effect is a reduction in the gas density due to adsorption, whereupon the density of the remanent gas decreases in the interval of pressures at which emission takes place.

§2. Emission at Ar III Transitions in the Ultraviolet Spectral Region

Inasmuch as the temperature and degree of ionization in the pinch effect increase abruptly at the instant of cumulative discharge, it may be assumed that negative-temperature states occur in the axial part of the discharge in the level systems of ions with a high ionization multiplicity. We observed emission at the 3511 Å Ar III transition ($4p^3P_2 \rightarrow 4s^3S_1^0$) and at the 3638 Å line corresponding either to Ar III or to argon with a high ionization multiplicity. The gain at the 3511 Å line is so great that emission occurs even with mirrors designed for the visible part of the spectrum and invested with a reflectivity of 65% in the emission interval. The gain at the 3637.9 Å line is much smaller, and emission occurs only when the reflectivity of one of the mirrors is raised to 99%. The emission at these lines was investigated in 8- to 20-mm-bore tubes, although it is observed in large-bore tubes as well. However, the pressure interval in which emission is observed is narrowed down to such an extent that it is exceedingly difficult to establish the proper initial pressure, and this, of course, creates experimental difficulties. Below we describe certain characteristics of the emission at the 3511 and 3637.9 Å lines and give a comparison with the corresponding characteristics of the emission at the 4764.89 Å line.

a. Inception and Duration of Emission. The individual emission lines were recorded with a monochromator and FÉU-14B photomultiplier. As before, the discharge current was recorded simultaneously. An oscillogram of the laser emission and current pulses is shown in Fig. 20. Also shown in the same figure for comparison is the emission pulse for the 4764.89 Å line. The emission at 3511 and 3637.9 Å begins 0.1 to 0.2 μsec later and therefore corresponds to the instant of cumulative discharge, at which time the emission at 4764.89 Å has almost cut off. The duration of emission at the ultraviolet lines is also approximately 0.2 μsec. The concurrence of cumulative discharge and emission at the 3511 and 3638 Å lines is also evinced by the fact that the diameter of the emission spot almost exactly corresponds to the diameter of the plasma filament at the instant of cumulative discharge. The temporal and spatial correlation of lasing at the different lines is clearly seen in Table 5. The data of this table

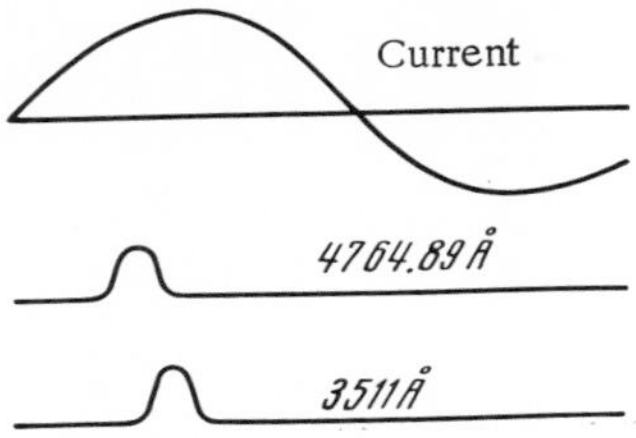

Fig. 20. Oscillograms of emission and current pulses in a 20-mm-bore tube (C = 0.1 μF; U = 25 kV).

TABLE 5

Emission wave-length, Å	Emission delay time relative to current start, μsec	Cumulative discharge time, μsec	Emission spot diameter, mm	Filament diameter at emission start, mm
3511	0.3–0.35	0.3–0.35	4–5	3
3637.9				
4764.89	0.2–0.25		12	10

correspond to the following discharge parameters: chamber diameter, 20 mm; initial pressure, $1 \cdot 10^{-2}$ torr; C = 0.1 μF; U = 25 kV.

b. Initial-Pressure Dependence of Emission Intensity. The nature of the dependence of the emission intensity on the initial pressure is the same as for emission in the visible part of the spectrum. For small-bore tubes the interval of optimum pressures is fairly broad and shifted slightly toward pressures larger than for the 4764.89 Å line. With an increase in the bore of the discharge chamber the size of the shift decreases, vanishing almost altogether for chambers with bore ≥ 15 mm. However, the width of the pressure interval shrinks to a much greater degree than for emission in the visible region. The corresponding curves for 8- and 20-mm-bore tubes are shown in Fig. 21. The optimum pressure intervals for the 3511 and 3637.9 Å lines coincide (Fig. 22). The intensity of the 3511 Å line is considerably greater than that of the 3637.9 Å line. For a 15-mm-bore tube the intensity ratio of these lines is 4:1. With an increase in bore this ratio decreases (2:1 for the 20-mm-bore tube). However, the absolute value of the intensity for each line is sharply increased in this case. The relative value of the intensities of the Ar II and Ar III emission lines was determined with an FÉU-18A photomultiplier, whose sensitivity is roughly equal in the near ultraviolet and blue regions of the spectrum. Attenuating light filters calibrated with a Japanese Shimadzu spectro-

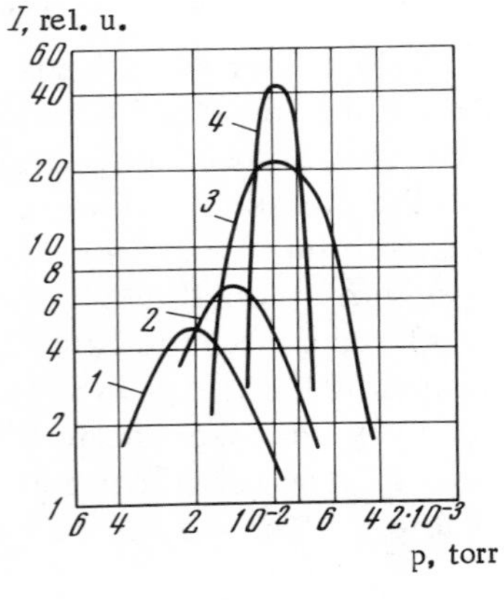

Fig. 21. Curves of emission intensity at the 4765 Å (2, 3) and 3511 Å (1, 4) lines in 8-mm (1, 2) and 20-mm (3, 4) tubes.

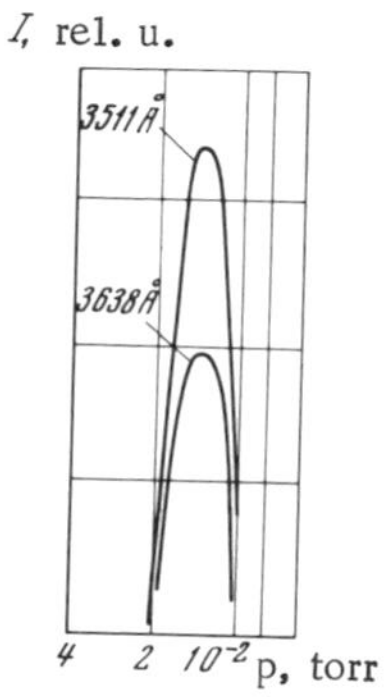

Fig. 22. Intensity of Ar III emission lines in a 20-mm-bore tube (C = 0.1 μF; U = 25 kV).

photometer were set up in front of the photomultiplier. It was ascertained that the emission intensity in the ultraviolet region is two or three times the visible intensity. Consequently, the total emission intensity in the 20-mm tube is about 1 kV.

§3. Emission in Subsequent Discharge Current Half-Periods

In several experiments two emission pulses were observed with an interim spacing of 1.2 to 3 μsec. This situation occurred in 8- to 20-mm-bore discharge tubes with the use of a 0.01-μF capacitor. The second emission pulse was observed in a relatively narrow interval of initial pressures somewhat above the optimum and in a narrow voltage interval: 15 to 20 kV. The period of the discharge current is equal to 0.8 μsec, and the peak current in 2 kA. The parameters of the optical cavity for observation of a second pulse were the same as in the experiments described above (80 and 99% reflectivity for the 4764.89 Å line and 65 and 70% in the vicinity of 3500 Å). Under these conditions emission was observed both at the 4764.89 Å line and at the 3511 Å line. In both cases the width of both the first and second emission pulse was 0.2 μsec at the half-peak intensity level. As in the one-shot case, emission in the visible region was almost always observed at just one line. The intensity of the subsequent pulses is the same (Fig. 23). The time interval between emission pulses in tubes of different diameters is the same and changes from one experiment to the next, remaining at all times a multiple of the half-period.

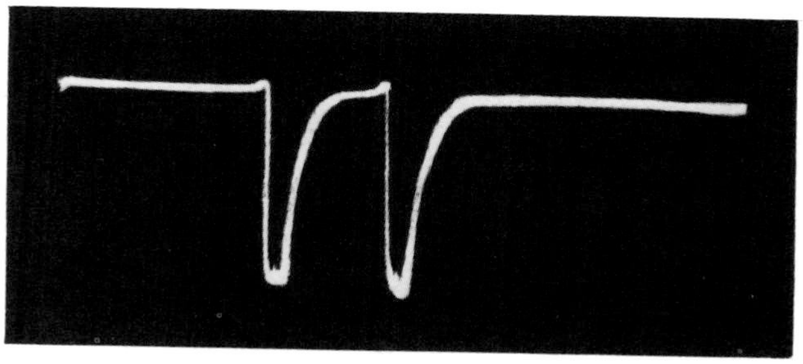

Fig. 23. Emission pulses in two discharge current periods in a 20-mm-bore chamber (C = 0.01 μF; U = 20 kV; p = $1.5 \cdot 10^{-3}$ torr; λ = 4764.89 Å: sweep time, 10 μsec).

Fig. 24. Photograph of pinch discharge in a 20-mm-bore chamber ($C = 0.01\,\mu F$; $U = 20$ kV; $p = 1.5 \cdot 10^{-2}$ torr).

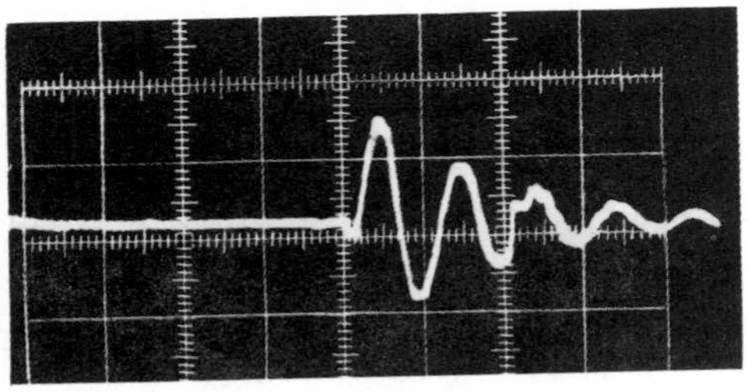

Fig. 25. Oscillogram of current in a 20-mm-bore chamber ($C = 0.01\,\mu F$; $U = 20$ kV; $p = 1.5 \cdot 10^{-2}$ torr; 1 scale division = 1 μsec).

The investigation of the second-emission regime was not included as part of the present program. We point out, however, that for the given discharge parameters complete pinch does not occur, because the current rise time is not sufficient. Nevertheless, the first phase of discharge development, i.e., detachment from the chamber wall and partial constriction, is the same as in complete pinch (Fig. 24). In the later discharge stages instabilities set in, corresponding to the filament luminescence fluctuations disclosed by high-speed photoscanning in the frame-by-frame mode. The onset of instabilities and fluctuations of the plasma filament are also indicated by the fluctuations of the discharge current (Fig. 25). The intensity of the fluctuations and, hence, of the laser-level excitation is modulated by the discharge current, thus accounting for the fact that the time interval between emission pulses is a multiple of the discharge current period. A highly detailed study of the time variation of the plasma filament configuration in discharges with approximately equivalent parameters is described in [69].

Double-pulse lasing has also been observed in [33, 34] (see Chapter I). In these papers, however, as opposed to our own investigation, the second emission pulse occurred either in the afterglow or at the end of the discharge current and corresponded to the 4879.9 Å line, whereas the first pulse corresponded to the 5764.89 Å line, indicating a difference in the excitation mechanisms.

§4. Emission at the 4764.89 Å Line for Discharge Current Pulses Having a High Repetition Rate

The power supply section used in all the experiments was designed to operate in the single current-pulse regime. However, during operation with a low capacitance (0.01 μF) in certain cases spontaneous breakdown occurred in the discharge chamber at a high repetition rate, corresponding to 100 kHz. The reason for this effect is understandable in light of the detailed supply circuit. The capacitor at the rectifier output (0.02 μF) is able for a certain period of time to boost the charge on the working capacitor. The boosting time is determined by the charge resistance (1000 Ω in our experiment) and is equal to 10^{-5} sec, i.e., corresponds to the breakdown repetition rate. The role of the nonlinear element required in any periodic signal generator is taken in our case by the spark gap. During the discharge period (4 μsec) the voltage on the capacitor drops to a value insufficient to sustain discharge. As a result, the current ceases until the capacitor builds up its charge to a voltage sufficient for a new breakdown. Up to ten successive breakdowns were usually observed. The time between adjacent current pulses was not the same, either in a single series of breakdowns or in different experiments.

In small-bore (8 mm) tubes emission generated in the successive current pulses was observed at the 4764.89 and 3511 Å lines, while in large-bore (20 mm) tubes it was observed only at the 4764.89 Å line. In the latter case a single emission occurs only in the first "shot." The emission amplitude in the successive current pulses is approximately identical, with a duration of 0.2 μsec, as in the ordinary experiments (Fig. 26).

For a more detailed investigation of the double-pulse emission regime we slightly modified the experimental arrangement (Fig. 27). Two successive current pulses were generated by two independent supply circuits consisting of a capacitor and spark gap fired by a special circuit with a variable (0.5 to 1000 μsec) delay time. The experiments were carried out in 8- and 20-mm-bore chambers under optimum lasing conditions. The emission was investigated only in the visible part of the spectrum. With the use of a 0.01-μF capacitor the second emission pulse remains equal to the first as the current pulses are brought within 8 μsec of one another in the 20-mm-bore tube and within 6 μsec in the 10-mm-bore tube. With the current pulses in closer proximity the amplitude of the second pulses decreases. When the width of the current pulses is diminished by the inclusion of a 10-Ω resistance in the discharge circuit the emission pulses can be spaced within 4 μsec of each other. With the use of a 0.1-μF capacitance the emission pulses can only be brought within 20 to 30 μsec apart without a loss of intensity of the second pulse. For a qualitative interpretation of these results the temperature was measured at the times corresponding to the second emission pulse. The temperature was determined from the relative intensity of the hydrogen H_α and H_β lines. These measurements show that the second emission pulse occurs only when the plasma is cooled to ~ 0.5 eV. The

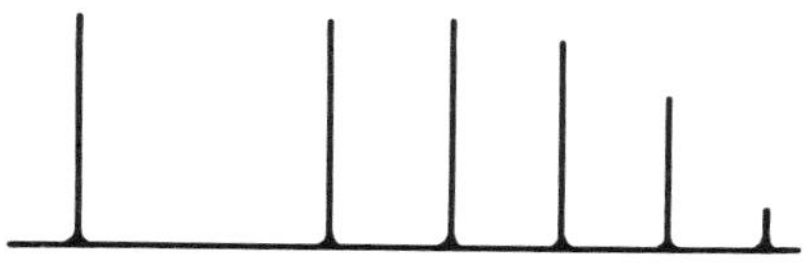

Fig. 26. Emission pulses at a high current-pulse repetition rate (λ = 4764.89 Å; C = 0.01 μF; U = 20 kv; p = $2 \cdot 10^{-2}$ torr; bore = 8 mm; sweep time, 50 μsec).

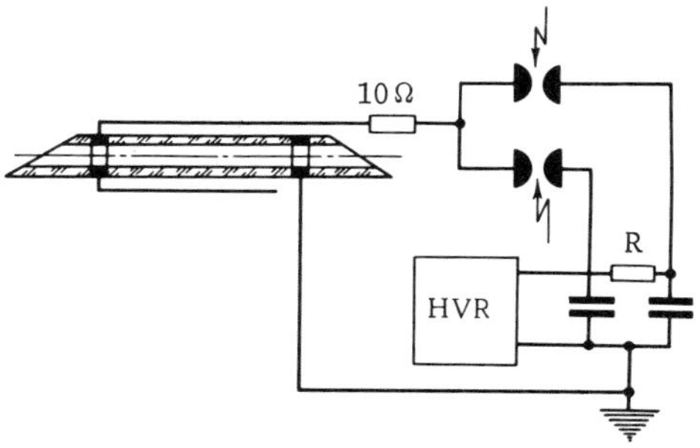

Fig. 27. Schematic of the system for investigation of the double-pulse emission regime.

disparity in the time intervals between emission pulses for different conditions is determined by the difference in the current-pulse durations (using different capacitors) and the disparate roles of diffusion at the wall for different discharge chamber diameters. Consequently, the experiments described here attest to the feasibility of a high-frequency lasing regime emitting at several hundred kilohertz. This regime is possible, of course, only in short pulse trains such that overheating of the system does not occur.

CHAPTER V

DETERMINATION OF THE PARAMETERS OF THE PINCH-DISCHARGE PLASMA

In order to fully understand the processes described in the preceding chapter one must have very detailed information on the temporal and spatial variations of the parameters of the discharge plasma. The times characterizing the development of the processes involved in pinch are on the order of 10^{-8} to 10^{-7} sec. To perform measurements with a commensurate resolution poses a very complex experimental problem. In addition to the experimental problems there are fundamental difficulties associated with the impracticality of many of the methods based on the presumed existence of thermodynamic equilibrium. The occurrence of strong deviations from thermodynamic equilibrium in a pinch-discharge plasma is already indicated by the mere observation of induced emission. The use of any method in this case requires special substantiation. This requirement is especially true of spectral techniques for determining the electron temperature. Spectral methods, however, permit a reliable determination of other parameters of the plasma (the ion temperature and ion and electron densities) without imposing stringent demands for the presence of thermodynamic equilibrium. An exceedingly important aspect of spectral methods is that they do not introduce perturbations into the investigated medium. Measurements of the electron temperature according to the plasma conductivity are the most reliable at high degrees of ionization. We now discuss the applicability of various plasma diagnostic methods under our experimental conditions.

§1. Theoretical Foundation of the Applicability of Plasma Diagnostic Methods

A. Determination of the Electron Temperature from the Plasma Conductivity. In general the conductivity of a plasma is a very complex function of both its temperature and density. The specific form of the expression for the plasma conductivity under diverse conditions has been studied in detail in [71]. The problem is most complex in the case of low degrees of ionization. This case requires the simultaneous analysis of electron scattering by ions and neutrals. As shown in Chapter II, the probabilities of these processes become identical for just a small percentage ionization. With an increase in the latter the role of Coulomb collisions becomes predominant. Clearly, it is entirely justified to neglect the contribution of scattering by neutrals for more than 50% ionization. In this event the conductivity depends only on the electron temperature and is determined by the well-known equation

$$\sigma = 1.16\left(\frac{2kT_e}{m\pi}\right)^{3/2}\frac{m}{e^2\Lambda} \approx 1.1\cdot 10^{13} T^{3/2}\ \mathrm{eV}. \quad (1)$$

Consequently, for large degrees of ionization a measurement of the conductivity affords a reliable method of determining the electron temperature. Under pinch conditions this method can be used with certainty at the instant of cumulative discharge, which is characterized by a large temperature and degree of ionization, as well as at times immediately preceding and following cumulative discharge. In these cases, however, it is essential, of course, to exercise definite precautions.

B. Spectroscopic Methods of Determining the Parameters of a High-Temperature Plasma. a. Determination of the Electron Temperature. We note first of all that over the entire range of variation of the discharge parameters in our experiment we are dealing with an optically thin medium. The emission spectrum consists mainly of individual lines. The electron-temperature measurement techniques are based on a measurement of the absolute or relative intensities of the spectral lines. The individual line intensity is determined as follows:

$$I_{ki} = N_k A_{ki} h\nu_{ki}, \tag{2}$$

where N_k is the population of the k-th level, A_{ki} is the transition probability, and $h\nu_{ki}$ is the energy per quantum. Among the basic processes governing the filling of the energy levels and, hence, the line intensities we distinguish spontaneous emission and electron collisions of the first and second kind:

$$\int n_e v f(v) N_0 \sigma_{0k} dv = \int v e^{E_k/kT} N_k f(v) \sigma_{0k} \frac{g_0}{g_k} dv + \Sigma N_k A_{ki}, \tag{3}$$

where N_0 is the density of ground-state neutrals, σ is the electron-impact excitation cross section, g is the statistical weight, $f(v)$ is the electron velocity distribution function, and n_e is the electron density. The relative contribution of spontaneous emission and impacts of the second kind to depopulation of the energy levels will differ, depending on the electron density. We consider two extreme cases.

Small Electron Density. Spontaneous emission plays the decisive role in depopulation processes. As a result, the line intensity is determined by the excitation rate of the given level by electron collisions:

$$I_{ki} = h\nu_{ki} N_0 n_e \int\limits_{v_k}^{\infty} \sigma_{0k}(v) v f(v) dv. \tag{4}$$

If the electrons have a Maxwellian velocity distribution, the line intensity is determined as follows:

$$I_{ki} = 2\left(\frac{2kT_e}{\pi m}\right)^{1/2} N_0 n_e \sigma_{0k} h\nu_{ki} \left(1 + \frac{E_k}{kT_e}\right) \exp\left(-\frac{E_k}{kT_e}\right), \tag{5}$$

where E_k is the energy of the excited level, T_e is the electron temperature, and m is the electron mass. It is evident from the foregoing that the electron temperature can be measured, in general, either from the absolute intensity of the individual lines or from the relative line intensity. The difficulty with the indicated method lies, on the one hand, in the lack of sufficient detailed information on the degrees of excitation in most cases and, on the other, in the ever-present possibility of error due to the disregard of other processes, such as cascade transitions and excitation from intermediate levels. Moreover, for large populations stimulated emission can play a significant part.

Large Electron Density. In this case radiation processes play a minor part relative to electron impacts. The population of the energy levels is determined by the Boltzmann equation subject to the condition that the velocity distribution of the electrons is Maxwellian:

$$N_k = \frac{g_k}{g_0} N_0 e^{-E_k/kT_e}. \tag{6}$$

Thus, the intensities of the individual lines are independent of the excitation cross sections, being determined solely by the electron temperature:

$$I_k = h\nu_{ki} A_{ki} \frac{g_k}{g_0} N_0 e^{-E_k/kT_e}. \tag{7}$$

However, very large electron densities are required in order to establish a Boltzmann distribution over all levels including the ground state. Nevertheless, an equilibrium distribution can obtain between individual groups of levels, even for considerably lower concentrations. It is difficult in general to estimate the electron density required for partial equilibrium. Such estimates are available, however, for hydrogen and for the upper (hydrogenlike) levels of other atoms [72]. For hydrogen, with an increase in the principal quantum number n the electron-excitation cross sections increase as n^4, and the spontaneous-emission probabilities decrease as $n^{9/2}$. Thus, given an arbitrary electron density, for the upper levels, beginning with a certain n, the population is determined by electron collisions and conforms with the Boltzmann distribution. A detailed analysis by Griem [72] shows that the electron density required for partial equilibrium between levels with principal quantum numbers greater than a given n is determined as follows:

$$n_e \geqslant 7 \cdot 10^{18} \frac{Z^7}{n^{17/2}} \left(\frac{kT_e}{Z^2 J_H}\right)^{1/2} \text{cm}^{-3}, \tag{8}$$

where Z is the charge of the atom minus an optical electron and J_H is the ionization potential of hydrogen. For a given density the equilibration time is determined by the relation

$$\tau_n = \frac{4{,}5 \cdot 10^{17} Z^3}{n^4 n_e} \left(\frac{kT_e}{Z^2 J_H}\right)^{1/2} \exp\left(\frac{E_{n'} - E_n}{kT_e}\right), \tag{9}$$

in which

$$E_{n'} - E_n = Z^2 J_H \left(\frac{1}{n'^2} - \frac{1}{n^2}\right) \tag{10}$$

is the energy difference between levels n' and n. Estimates according to these equations show that in hydrogen for $n \geq 3$ a Boltzmann distribution is established for $n_e \sim 5 \cdot 10^{14}$ cm over a wide temperature range. The equilibration time at these densities does not exceed 10^{-9} sec. It may be assumed, therefore, that even in a pinch discharge a Boltzmann distribution can be established for hydrogen. The lines of the Balmer series H_α and H_β can be used to measure the temperature. This can be done by injecting hydrogen into the discharge as a small impurity not affecting the discharge and lasing conditions. The temperature is determined from the line intensity ratio for a Boltzmann distribution as follows:

$$T_e = \frac{0.43\,(E_2 - E_1)}{k\left\{\log \frac{J_1}{J_2} - \log\left[(\nu_1/\nu_2)^3\,(g_1 f_1/g_2 f_2)\right]\right\}}. \tag{11}$$

The quantities g_1, g_2, f_1, and f_2 are known for hydrogen. Substituting them into (11), we finally deduce as our working equation

$$T_e = \frac{0.26}{\log(J_{H_\alpha}/J_{H_\beta}) - 0.06} \text{ eV}. \quad (12)$$

b. Determination of the Ion Temperature. For sufficiently large ion temperatures the Doppler broadening of the lines can be used for measurements. Given a Maxwellian ion velocity distribution, the line has a Gaussian contour, and its half-width is equal to

$$\Delta\lambda_D = 7 \cdot 10^{-7} \lambda_0 \sqrt{\frac{T_i}{M}}, \quad (13)$$

where T_i is the ion temperature, M is the atomic weight, and λ_0 is the wavelength of the line. We note that a Maxwellian distribution is established relatively quickly [54]:

$$t_M = \frac{11.4 A^{1/3} T_i^{3/2}}{n Z^4 \ln \Lambda}. \quad (14)$$

The Maxwellian randomization time determined for the time of cumulative discharge under our experimental conditions is $\sim 10^{-7}$ sec. This time is therefore commensurate with the holding time of the plasma filament in the pinched state. In addition to deviations from a Maxwellian distribution, the line contour can also be distorted by directional ion flows. The temperature measurements should therefore be regarded as satisfactory only when the line contour has a nearly Gaussian configuration and the temperatures determined from the lines for ions having different ionization multiplicities coincide.

c. Determination of the Electron and Ion Densities. The electron and ion densities can be determined from the linear Stark effect for hydrogen and hydrogenlike ions and from the square-law effect for all other atoms. The square-law Stark effect leads both to broadening and to displacement of the spectral lines, as determined by the following relations:

$$\Delta\lambda = 9.8 C_4^{2/3} v_e^{1/3} n_e \left[I'(\beta) + \left(\frac{2mZ^2}{M}\right)^{1/6} \right],$$

$$\delta\lambda = 9.8 C_4^{2/3} v_e^{1/3} n_e \left[I''(\beta) + \left(\frac{2mZ}{M}\right)^{1/6} a^{1/4} \right], \quad (15)$$

in which C_4 is the square-law Stark constant and the terms $I'(\beta)$ and $I''(\beta)$ are introduced in the nonstationary theory of the square-law Stark effect. The constants C_4 are small, so that for small electron densities, $\sim 10^{15}$ cm^{-3}, high-resolution spectral equipment is required to record the broadening. The constants of the linear Stark effect are considerably greater, and the broadening of, for example, the hydrogen lines at such densities attains several angstroms. At large temperatures the thermal motion of both electrons and ions will contribute to the broadening. A rigorous calculation by Kogan for this case [73] gives the following relation between the line width and ion density:

$$8 \cdot 10^{11} \Delta\lambda = 23 C_2 N_i^{2/3}, \quad (16)$$

in which C_2 is the linear Stark constant and $\Delta\lambda$ is the total line half-width in angstroms. Normally the H_β line, for which $C_2 = 10.5$ cm^3, is used for the measurements. We note that the velocity dependence of the Stark broadening is weak and the given equations hold even with strong deviations from a Maxwellian distribution.

§2. Measurement Procedure and Results

A schematic diagram of the measurement apparatus used for the present plasma-diagnostic studies is shown in Fig. 28. In every case, unless specifically stated to the contrary, the measurements were performed for a 20-mm-bore tube at the optimum pressure for lasing (see also [74]). We used 0.1- to 0.4-μF capacitors at a voltage of 25 kV. The spectrum was recorded with an STÉ-1 spectrograph. KN-13 or RF-3 high-sensitivity aerial photography film was used for the measurements. The line contours were recorded on an MF-4 photomicrometer. A monochromator (ZMR-3) with a photoelectric attachment was used for time scanning of the individual line intensities. The duration of the emission for the individual lines of the working gas, as indicated by the oscillograms (Fig. 29), is equal to 0.5 to 0.8 μsec for various capacitances. The peak intensity corresponds to the instant of cumulative discharge. Thus, several of the plasma parameters at the time of cumulative discharge can be determined from ordinary spectrograms without resorting to time scanning of the spectrum. The values thus found were averaged over a relatively short time interval.

Determination of the Ion Temperature. The ion temperature was determined from the width of the Ar II lines (4806.0, 4847.8 Å, etc.) with small square-law Stark constants ($C_4 = 5 \cdot 10^{-15}$ cm^4 sec^{-1}) and from the Ar III lines (3480.5, 3514.18 Å). For ion densities of 10^{16} cm^{-3} or lower it may be assumed that the Stark broadening is already considerably less than the Doppler broadening at ion temperatures of order 1 eV. Inasmuch as high-sensitivity film was used in the experiment, it was the main factor responsible for instrumental broadening. The total instrument characteristic of the measurement system (including the microphotometer) was determined from photomicrograms of the lines of iron or a standard helium–neon laser. The instrument characteristic was nearly Gaussian. Since the Doppler contour is also Gaussian, its half-width is readily determined:

$$\Delta\lambda_D = \sqrt{(\Delta\Lambda)^2 - (\Delta\lambda_{app})^2},$$

where $\Delta\Lambda$ is the total half-width of the line and $\Delta\lambda_{app}$ is the half-width due to the apparatus

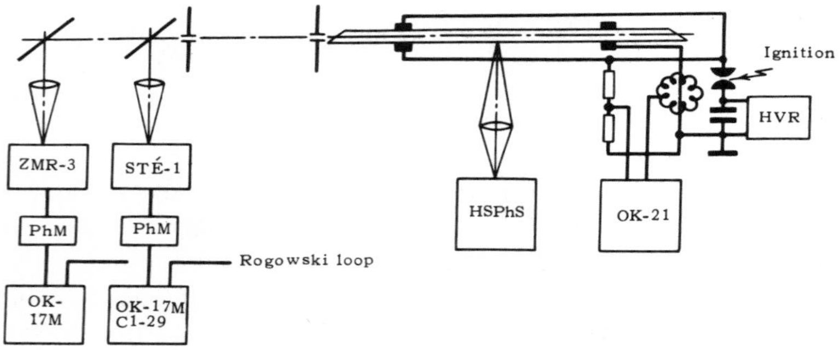

Fig. 28. Schematic of measurement apparatus used for plasma diagnostics.

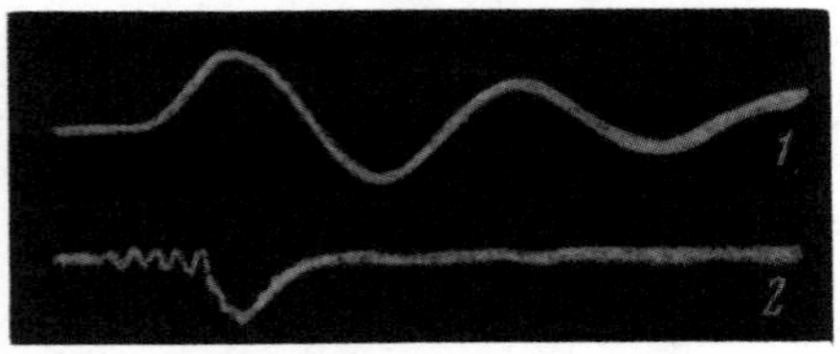

Fig. 29. Oscillogram of current (1) and emission at the Ar II 4806 Å line (2) (bore = 20 mm; C = 0.1 μF; U = 25 kV; p = 10^{-2} torr; sweep time, 6 μsec).

TABLE 6

	Wavelength, Å	T, eV ($C = 0.1\ \mu F$)	T, eV ($C = 0.4\ \mu F$)
ArII	4806.0	32	53
	4847.8	36	58
ArIII	3480.5	33	55
	3514.18	38	60

(instrument half-width). The half-widths of the Ar II and Ar III lines after elimination of the instrument half-width are 0.23 to 0.45 Å for various capacitances. The temperatures determined from the various lines coincide for each capacitance. The temperatures determined from some of the lines are given in Table 6.

The coincidence of the temperature determined from the lines for ions having different ionization multiplicities attests to the reliability of the values obtained. The increase in temperature with the capacitance is explained by the fact that the holding time of the plasma filament in the pinched state, as indicated by high-speed photoscanning (HSPhS) exposures, is greater for large capacitances, so that larger current densities and temperature are attained. Also, the relative proportion of emission corresponding to the high-temperature medium increases in the aggregate. We note that emission at Ar III occurs for the smaller of the given temperatures. This result may be inferred from the fact that an increase in the capacitance above 0.1 μF does not affect the nature of the emission, while the increase in temperature with the capacitance happens on the falling side of the emission intensity contour.

Determination of the Electron and Ion Densities. We attempted in the beginning to determine the electron density from the working-gas lines broadened due to the square-law Stark effect. For this purpose we measured the width of the majority of Ar II lines in the 3500 to 6000 Å interval, which have the largest Stark constants. It turned out that the widths of these lines are only slightly greater than the Doppler line width. Consequently, in order to isolate the Stark width it is necessary to separate the Stark and Doppler contours. The dispersion of the spectral instruments at our disposal was not good enough for the scrupulous execution of this operation. Measurements of this nature were therefore not performed. However, rough estimates of the width and shift of the spectral line (0.1 ± 0.05 Å) afforded upper-bound estimates of the densities in our experiment. It follows from such estimates that the electron density does not exceed 10^{16} cm^{-3} at the instant of cumulative discharge. The measurement of the widths of the hydrogen lines broadened by the square-law Stark effect is more reliable and simpler in the interval of densities below 10^{16} cm^{-3}. Under our experimental conditions the hydrogen lines were generally present in the discharge spectrum, but their detection (in single discharges) was impossible except by photoelectric techniques. When such detection became too difficult, a small hydrogen impurity (a few percent) was injected into the discharge, where it did not affect the emission properties. The large line width makes it possible to perform time scanning of the line and, hence, of the density. We used the following procedure. Separate subintervals of the contour of the H_β line were partitioned off by means of a micrometer slit set up in the focal plane of the STÉ-1 spectrograph. The slit displacement error was 0.01 mm, which corresponds to 0.06 Å in the vicinity of 4861 Å (H_β). Thus, with fairly good reproducibility of the measurements the contour of the line can be plotted for each instant from the resulting curves. This was accomplished by recording from the end of the discharge chamber, so that the data scatter due to plasma density fluctuations along the length of the discharge, which are inescapable in recording from the side, tend to be smoothed out. For the measurements the emission from the central part of the discharge, which was 4 mm in diameter, was carefully separated out. In the recording of emission from the end of the chamber there is the danger of error associated with resorption, which causes the line width and, hence, the density to be recorded too large. The absence of resorption in our experiment was confirmed by varying the hydrogen concentration by a factor of five to ten. The fact that the half-width of the H_α and H_β lines remained constant attests to the absence of appreciable resorption. The line contours plotted for different measurement series coincide correct to 10%. The line half-width at the time of cumulative discharge is 9.4 Å. At the ion temperatures given above, however, the Doppler effect makes a sizable contribution to this width. In order to include the Doppler broadening, in general, it is impossible to use the ion temperature determined from the ionized-argon lines, because the peaks of the hydrogen and argon line intensities can be shifted along the diameter of the plasma filament and correspond to different temperatures. For a more precise determination of the Doppler broadening contribution, therefore, we also performed time scanning of the contour of the H_α line. The relation between the Stark and Doppler widths of the H_α and H_β lines in the relevant concentration interval is given in [75]; this relation enables us to write a simple set of equations, from which the Doppler width is easily found:

$$\Delta\lambda_\alpha = \Delta\lambda_{D\alpha} + \Delta\lambda_{S\alpha}, \qquad \Delta\lambda_{D\alpha} = k\Delta\lambda_{D\beta},$$
$$\Delta\lambda_\beta = \Delta\lambda_{D\beta} + \Delta\lambda_{S\beta}, \qquad \Delta\lambda_{S\alpha} = k\Delta\lambda_{D\beta}.$$

At the time of cumulative discharge the half-widths of the H_α and H_β lines are equal to 1.5 and 9.4 Å, respectively. Consequently, the Doppler width of the H_β line is 1.9 Å. The temperature corresponding to this width is 15 eV, i.e., is considerably below the ion temperature determined for argon, a result that is consistent with the foregoing considerations. We note that practically all the hydrogen must be ionized in an equilibrium plasma at a temperature of 15 eV. The observation of the, albeit weak, hydrogen lines in our experiment is attributable to the fact that ionization equilibrium does not have time to be established in the short life span of the pinched filament. The Stark width of the H_β line was found by simply subtracting the Doppler component from the total line width; it turns out to be equal to 7.5 Å. The ion density determined according to Eq. (16) is equal to $4 \cdot 10^{15}$ cm^{-3}. The electron density can be found from the ion density by multiplying the latter by the effective charge and may be assumed equal to (6 to 8)$\cdot 10^{15}$ cm^{-3}. Consequently, the plasma density at the instant of cumulative discharge exceeds the initial density by approximately one order of magnitude. But the volume occupied by the plasma at that instant, as determined from HSPhS exposures, is approximately two orders of magnitude smaller. The implication is that only 10 to 20% of the gas is involved in the pinch effect.

The plasma density corresponding to subsequent pinches is approximately the same as in the first pinch. The resolution of the instrumentation was not adequate to allow the required measurements of the line width at times preceding cumulative discharge.

Measurement of the Electron Temperature from the Plasma Conductivity. The plasma conductivity was determined by simultaneously recording the discharge current and voltage drop across the discharge gap (see Chapter II). The voltage drop is determined as follows:

$$U = JR + L\dot{J} + J\dot{L}.$$

Knowing the discharge geometry as determined by high-speed photoscanning, one can take account of the role of the last two terms and separate the active component of the voltage drop. The active component of the voltage and, hence, the conductivity and electron temperature are determined with the least error at the time of cumulative discharge, when $J\dot{L} = 0$. Moreover, for a capacitance of 0.1 μF the term $L\dot{J}$ is also small at this time. Due to the large degree of ionization at the time of cumulative discharge the conductivity equation (1) is satisfied with the required safety margin. The temperature determined from this equation at the time of cumulative discharge is 14 ± 5 eV, and after 0.5 μsec the temperature decreases to 3 eV (Fig. 30).

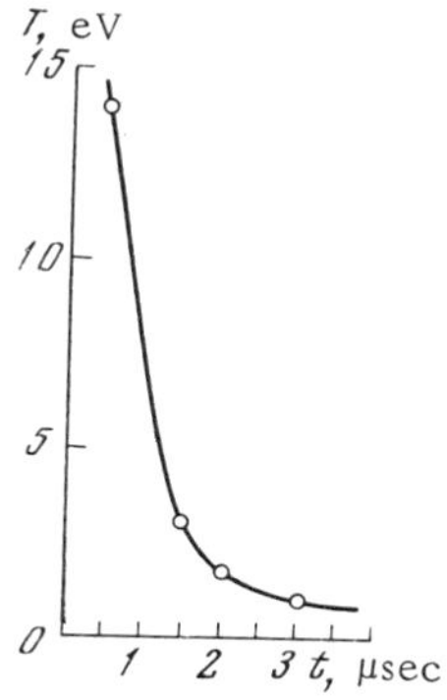

Fig. 30. Graph of T_e at the center of the discharge chamber (C = 0.1 μF; U = 25 kV; p = 10^{-2} torr; bore = 20 mm).

The electron temperature thus determined for a tube 20 mm in diameter at the time of the cumulative process is equal to 25 ± 5 eV.

Determination of the Electron Temperature from the Relative Intensity of the Hydrogen H_α and H_β Lines. Measurements based on the relative intensity of the spectral lines have the greatest accuracy in the range of temperatures commensurate with the energy difference between the levels from which emission originates. For the hydrogen H_α and H_β lines this difference is 0.6 eV. Consequently, hydrogen measurements, which allow the measurement of small temperatures [when the degree of ionization is too small for Eq. (1) relating the conductivity and temperature to be valid], complement the plasma conductivity measurements. Moreover, there is a temperature interval in which both methods can be used.

A monochromator with a photoelectric accessory was used to record the individual lines. In order to ensure good reproducibility, as in the case of the density measurements, the emission was recorded from the end of the discharge chamber. An FÉU-14B photomultiplier calibrated in the vicinity of the H_α and H_β wavelengths (6563 and 4861 Å) was used for the measurements. The relevant time was fixed by simultaneous recording of the discharge current. On account of the high sensitivity of the photomultiplier in the blue region of the spectrum, both lines yielded an approximately identical signal at the output, thus eliminating the errors associated with calibration of the oscilloscope amplifier. A curve showing the electron temperature variation in the center of the chamber is given in Fig. 30. The first point was obtained only from the conductivity measurements.

§3. Feasibility of Plasma Diagnostics Based on the Self-Induced Emission of a Pinch Discharge

a. Possible Mechanisms Responsible for the Shift of the Emission Wavelength. Due to the high temperature and plasma density in a pinch discharge the properties of the emission must be affected bý various intrinsic plasma effects. This statement is particularly true of emission by multiple ions. The shift of the emission wavelength can be attributed to at least two mechanisms. First, for large enough densities the Stark shift can be appreciable. Second, a shift can occur due to nonlinearity instability in a plasma containing inverted-population ions [76]. This instability is determined by the nonlinear interaction of a plasma wave, which can be generated in the plasma under definite conditions, with the optical wave at the laser-transition frequency. In this case nonlinear instability develops more rapidly than exponential by virtue of the fact that the characteristic increment of the instability is an increasing time function. Inasmuch as plasma oscillations can be excited rather rapidly in a time much smaller than the energy level lifetimes, they provide a possible mechanism of rapid de-excitation of the system. The characteristic time for the growth of instability and depopulation of the upper laser level is the plasma wave period. The emission wavelength must be shifted relative to the wavelength of the unperturbed laser transition by an amount determined by the plasma frequency

$$\omega_p = \left(\frac{4\pi n_e e^2}{m_e}\right)^{1/2}.$$

This shift could also be used to determine the electron density at the exact instant of emission. In our experiment, with the given effect present, one should expect an emission wavelength shift equal to ~1 Å. The measurements showed, however, that none of the emission lines was shifted beyond the experimental error; 0.05 Å. We should not dismiss the possibility of observing the effect described here in more powerful plasma devices.

b. Possibility of Plasma Diagnostics Based on Incoherent Scattering of the Self-Induced Emission. In recent years laser techniques have become commonplace in plasma diagnostics. We now examine some of them. For plasma interferometry with the use of helium-neon [77] or ruby (e.g., [78]) lasers the dielectric constant of the plasma is determined. This method is used to determine the time and space variation of the electron density. The time variation of the population of certain energy levels is determined from the coherent scattering of a laser beam. Obviously, in this case there must be close agreement of the energy of the laser transition and of the transitions corresponding to the investigated levels. This method has been used in [79] to determine the populations of the Ar II lines in a θ-pinch plasma in terms of the scattering of light from a continuous argon laser. The most widely used laser diagnostic method is the use of incoherent scattering of a ruby laser beam by fluctuations of the electron density or by free electrons. This method is endowed with the greatest complexity due to the smallness of the scattering cross sections. However, in principle it does permit one to obtain the most complete information about a plasma, namely the temperature and density of the electron and ion components. Surveys of laser diagnostic techniques may be found in [80, 81]. Here we consider only certain results of the theory. The nature of the scattering is determined by the ratio of the emission wavelength to the Debye radius:

$$\alpha = \frac{\lambda_0}{\lambda_D \sin(\theta/2)},$$

where λ_0 is the laser wavelength, λ_D is the Debye radius for electrons, and θ is the scattering angle. The scattering spectrum has a different character, depending on the value of the parameter α:

$\alpha \ll 1$: This condition implies that scattering takes place at individual electrons (ordinary Thomson scattering). The effective scattering cross section is determined by the classical electron radius:

$$\pi r_0^2 = \frac{\pi e^4}{m^2 c^4} \approx \pi \cdot 10^{-26}.$$

For a Maxwellian electron velocity distribution the contour of the scattering line is Gaussian with center at the incident-light wavelength. The half-width of the line is

$$\Delta\lambda = 1.7\lambda_0 \sqrt{\frac{kT_e \cdot 2}{mc^2}}.$$

For $\theta = 90°$ and $\lambda_0 = 6943$ Å we find $\Delta\lambda = 0.3\ [T_e(°K)]^{1/2}$. Therefore, the electron temperature is determined directly from the width of the contour. The electron density can be determined from the absolute intensity of the scattered light.

$\alpha \gg 1$: This second extreme case implies that scattering takes place at electron density fluctuations. The scattering spectrum consists of a central line at the emission wavelength and two satellites situated on either side of the central line at a distance determined by the plasma frequency ω_p:

$$\Delta\lambda = \frac{\omega_p}{2\pi c}\lambda_0^2\left(1 + \frac{3}{\alpha^2}\right)^{1/2}.$$

The width of the central peak is determined by the ion thermal velocities and makes it possible to find the ion temperature and, from the shape of the line contour, the ratio of the ion and electron temperatures.

$\alpha \sim 1$: The contour of the scattering line has an intermediate form. The configuration of the contour has been calculated for various values of α and electron-ion temperature ratios. Thus, the fundamental plasma parameters are determined from the scattering in this case. We note that all three cases considered here can almost always be realized in a single experiment by varying the angle of observation.

The scattering of a laser beam with a sufficiently large power and radiation energy can also be observed in principle in the laser itself (plasma-type). In this case, in addition to the incoherent variety, a considerably stronger coherent scattering is observed at the laser transition. However, the spectral composition of the scattered light differs (the width of the coherent scattering line does not exceed the emission line width), so that it is conceptually possible to observe the two effects in isolation. Estimates of the laser energy required in order to record the scattered light at various electron densities are given in [80]. For example, at an energy of 10^{-3} J the electron density must be at least 10^{15} cm^{-3}. Consequently, even for the relatively small n_e values in our experiment the scattering effect is detectable. Measurements of this type should be the object of an independent study and were not performed in the present investigation.

DISCUSSION

The experiments described in the preceding chapters afford a qualitative picture of the mechanisms for the excitation of negative-temperature states in a pinch-discharge plasma. The processes involved in the various stages of discharge development and the corresponding properties of the emission have significantly distinct characters. We propose to consider two stages of the discharge: the constriction process, or pinch effect, in the last phase of which lasing occurs at Ar II ions; and the cumulative discharge, when lasing by Ar II ions stops and lasing is generated by ions having a higher ionization multiplicity. We consider these two stages separately.

Pinch Effect. During the constriction of the plasma sheath there is an increase in the current density and, hence, in the temperature and degree of ionization of the plasma. This result is indicated by the rapid increase in intensity of the Ar II lines as the plasma front approaches the center of the chamber (see Fig. 14). The increase in the electron temperature is also evinced by the increase in the plasma conductivity with time. We note that a large difference exists between the electron and ion temperatures in the constricting plasma sheath. The temperature equalization time in this case, as shown in Chapter II, is 10^{-5} sec, which is more than an order of magnitude greater than the pinch time. We also point out that the mean-free path in the constricting sheath at initial pressures optimum for lasing is commensurate with the radius of the discharge chamber. This fact indicates that the shock wave cannot have a significant part in the excitation of the upper laser levels under our experimental conditions.

Generally speaking, both direct excitation of the upper laser levels [10, 16] and step-by-step excitation [44] occur during pinch. However, as inferred from [85], for an excitation period smaller than 10^{-6} sec the mechanism of direct excitation from ground-state neutrals prevails, and the 4764.89 Å line must be the most intense, a conclusion that is completely in harmony with the results of our experiment. In the presence of intense step-by-step excitation the 4879.9 Å line should be observed, but it was consistently absent or extremely weak in our experiment. We therefore consider only direct excitation. The direct excitation of atoms moving along with the plasma sheath clearly does not contribute significantly to the creation of population inversion, on account of the large degree of ionization and relatively high temperature, as well as, therefore, the low density of ground-state neutrals. The most probable event is that the upper laser levels are filled due to the excitation of neutrals becoming entrapped in the moving plasma sheath. This assumption simplifies the analysis, which is reduced to accounting for the interaction of a moving plasma sheath having a definite thickness and the time-varying temperature and density. The excitation of the laser levels takes place in accordance with the cross sections given in [27]. The excitation cross section for the lower laser level for the 4764.89 Å line is inconsequential, and its filling due to electron collisions can be neglected. Since the lifetime of this level is almost an order of magnitude smaller than that of the upper laser level [17], it may be assumed that when lasing does not occur or is very weak the population of this level is insignificant. Thus, the population inversion and emission intensity are determined by the quantity

$$\frac{dN^*}{dt} = N_0 n_e \langle v\sigma \rangle,$$

where N^* is the population of the upper laser level; N_0 is the density of ground-state neutrals, which is equal to the initial density in the discharge chamber; v is the electron velocity, n_e is the electron density, σ is the electronic excitation cross section, and the angle brackets $\langle\ \rangle$ indicate averaging over the Maxwellian distribution. The quantity $\langle v\sigma \rangle$ depends very strongly on the temperature. The corresponding values calculated for different temperatures are given in Table 7.

In general, the entries in the table need to be multiplied by a factor of order unity in order to correct for cascade filling from levels above the laser level (see [24]). The temperature corresponding to the self-excitation condition is easily estimated from the cited data. The population inversion required for the inception of lasing (in our case simply the population of the upper laser level) is determined from the requirement that the gain exceed the losses in the cavity:

$$g_0 = \sqrt{\frac{\ln 2}{\pi}}\,\frac{A}{4\pi}\,N^*\,\frac{\lambda_0^2}{\Delta\nu_D}, \tag{1}$$

where g_0 is the gain per unit length to overcome the cavity losses and is equal to $\sim 10^{-4}$ cm^{-1}, A is the laser transition probability, λ_0 is the emission wavelength, and $\Delta\nu_D$ is the Doppler line width, which is equal to 10^9 sec^{-1} for argon at room temperature. For the 4764.89 Å line we

TABLE 7

T_e, eV	$\langle v\sigma \rangle$, sec^{-1} cm^{-1}	T_e, eV	$\langle v\sigma \rangle$, sec^{-1} cm^{-1}
2	$2.5 \cdot 10^{-21}$	6	$9 \cdot 10^{-14}$
3	$1.3 \cdot 10^{-17}$	7	$3.7 \cdot 10^{-13}$
4	$1.2 \cdot 10^{-15}$	8	$9.3 \cdot 10^{-13}$
5	$1.6 \cdot 10^{-14}$	9	$1.9 \cdot 10^{-12}$
		10	$4 \cdot 10^{-12}$

have $A = 5 \cdot 10^7$ sec^{-1}. Consequently, the quantity N* must be 10^7 cm^{-3}. The excitation rate needed to sustain this inversion density is determined from the balance equation, which has the following form in crude approximation:

$$\frac{N^*}{\tau} = N_0 n_e \langle v\sigma \rangle,$$

where τ is the laser level lifetime, which is equal to 10^{-8} sec for $4p^3P_{3/2}$ (4765 Å) (see Table 7). Hence $dN^*/dt = 10^{15}$ cm^{-3} sec^{-1}. Substituting for our conditions $n = 10^{14}$ cm^{-3} and $N_0 = 3 \cdot 10^{14}$ cm^{-3}, we find that $\langle v\sigma \rangle = 3 \cdot 10^{-14}$ cm^{-1} sec^{-1}. In Table 7 this value corresponds to an electron temperature of 5 or 6 eV, which is the lasing threshold.

The absence of measurements with adequate time resolution in the pinch effect prevents us, unfortunately, from making a strict interpretation of the results pertaining to emission at the 4764.89 Å line. Even qualitative considerations based on the given model, however, enable us to account for a number of the experimental dependences. Thus, the delay of emission relative to the initial of the discharge current is determined by the heating time of the electronic component to the threshold temperature.

With an increase in the diameter of the discharge chamber the threshold temperature and, hence, lasing occur later, because the requisite current density is reached later (for an identical current growth rate). The inception of lasing is attained more rapidly with an increase in the current growth rate. Thus, with a fourfold increase in $\dot{J}$ lasing is initiated about twice as fast. The lasing period is limited to the time from its initiation to the instant of cumulative discharge, when the degree of ionization rises sharply and the emission intensity is decreased. Logically, with increasing $\dot{J}$ the lasing period decreases.

The pressure dependence of the emission intensity can also be qualitatively explained. At pressures above the optimum the resorption of induced emission at the lower laser level begins to be manifest, as evinced by the dark spot in the center of the emission spot (see Fig. 19). Moreover, the emission intensity can decrease by virtue of the decrease in electron temperature at large pressures, resulting in a reduction of $\langle v\sigma \rangle$ and the excitation efficiency of the laser levels. The reduction in emission intensity with the pressure is governed simply by the decrease in density of active centers.

Cumulative Discharge. The most intense excitation processes occur in the cumulative discharge stage. This fact is explained by the following causes: a) The density and degree of ionization of the plasma climb sharply. Under our conditions the ion density increases approximately tenfold over the initial density of neutrals, while the electron density increases twenty times; b) the current density increases many times (by one or two orders of magnitude); c) finally, the most important consideration is the transfer of energy from the ionic to the electronic component and the resulting further increase in the electron temperature. The ion kinetic energy accumulated throughout the entire pinch time is very rapidly converted into thermal energy at the time of cumulative discharge. In this case the ion temperature in our experiment attains 40 to 50 eV (see Chapter V). The electron temperature is almost an order of magnitude smaller. The energy transfer process is described by the relation

$$\frac{dW}{dt} = \frac{1.2 \cdot 10^{-17}}{A} n_e^2 \frac{T_e - T_i}{T_e^{3/2}},$$

in which A is the atomic weight of the ion and T_i is the ion temperature. The temperature equalization time under the condition $v_e \gg v_i$ (which is always satisfied at the instant of cumulative discharge) is equal to

$$\tau = \frac{17 A T_e^{3/2}}{n_e}.$$

Substituting into these equations the values measured experimentally, we obtain a temperature equalization time $\tau = 10^{-6}$ to 10^{-7} sec, and $dW/dt = 10^{12}$ ergs/sec $\cdot$ cm^3. The electrons very rapidly lose energy due to inelastic collisions, so that after the cessation of energy exchange from the ionic component the electron temperature drops rapidly (see Fig. 30). The lasing period (associated with Ar III ions) is limited by two factors: the cessation of intense pumping after temperature equalization and disruption of the plasma filament, resulting in a decrease in the temperature and density of the medium.

The lack of data on the electron excitation cross sections and transition probabilities for the system of Ar III levels prevents us from ascertaining the specific mechanisms for excitation of the laser levels. Certain estimates can be made, however, on the basis of qualitative considerations. Let us assume that the energy transfer efficiency into the excitation channel of the upper laser level is determined by a certain quantity α, which depends on the specific quantum system and on the electron temperature. We can determine this quantity from the equation

$$\frac{dW}{dt} \alpha l \frac{\pi d^2}{4} = W_L,$$

where d is the diameter of the active medium, l is its length, and W_L is the radiated laser power. In our experiment d = 0.4 cm, l = 100 cm, and W_L = 1 kW. Consequently, $\alpha = 10^{-3}$. This quantity, clearly, can be increased by increasing T_e, i.e., by increasing the fraction of electrons with energies above the threshold for the excitation of the upper laser level. The power can also be increased by increasing the electron density, i.e., by increasing dW/dt. Thus, it is possible, evidently, to enhance the radiated power by one or two orders of magnitude. And in fact a power of 40 kW has been obtained for N III ions in [3]. Besides increasing the laser power, the large energy influx rate at the time of cumulative discharge offers considerable possibilities for lasing in the short-wave part of the spectrum, including the vacuum ultraviolet. As apparent from (1), the excitation rate of the upper laser level is determined as follows:

$$NA = \frac{4\pi}{\sqrt{\ln 2/\pi}} \frac{g_0 \Delta\nu_D}{\lambda_0^2}. \tag{2}$$

The pumping intensity must increase sharply with decreasing emission wavelength. This fact is attributable to the decrease in lifetime of the levels relative to the corresponding transitions. The situation is further complicated by the large threshold gains by comparison with the optical range. This result is accounted for by two facts. First, mirrors of adequate quality are not currently available for the vacuum ultraviolet region. Second, the demand for increased pumping power means that the pumping period must be decreased [$\sim (dW/dt)^{-1/2}$ in the pinch effect] to values commensurate with the transit time of the light pulse through the active medium, thus precluding, in principle, the possibility of using an optical cavity. Lasing can only be realized, therefore, in the superradiance state. The threshold gain in this case is at least two orders of magnitude higher. An estimate according to Eq. (2) for $\lambda_0 = 1500$ Å and $T_i = 50$ eV gives the excitation rate needed for superradiant lasing as equal to 10^{21} cm^{-3} sec^{-1}. Assuming that the excitation efficiency α, as in the Ar III case, is equal to 10^{-3}, we obtain the required energy influx rate (considering the excitation energy for the upper laser level to be 70 eV):

$$\frac{dW}{dt} = 10^{14} \text{ ergs/cm}^3\text{sec} .$$

This quantity is only two orders of magnitude greater than the energy influx rate in our experiment and is technically quite attainable.

LITERATURE CITED

1. S. G. Kulagin, V. M. Likhachev, E. V. Markuzon, et al., ZhÉTF Pis. Red., 3:12 (1966).
2. V. M. Likhachev, M. S. Rabinovich, and V. M. Sutovskii, Eighth Internat. Conf. Ionized Phenomena in Gases (1967), p. 259.
3. J. S. Hitt and W. T. Haswell, IEEE J. Quantum Electronics, QE-2(4):6C-4 (1966).
4. Yu. V. Tkach, Ya. B. Fainberg, L. I. Bolotin, et al., ZhÉTF Pis. Red., 6:956 (1967).
5. D. E. McCumber and D. E. Platzman, Phys. Fluids, 6:1446 (1963).
6. W. E. Bell, Appl. Phys. Lett., 4:34 (1964).
7. A. L. Bloom, W. E. Bell, and D. L. Hardwick, Bull. Am. Phys. Soc., 9:143 (1964).
8. M. Armand and P. Martinot-Lagarde, Compt. Rend., 258:867 (1964).
9. W. B. Bridges, Appl. Phys. Lett., 4:128 (1964).
10. W. R. Bennett, Jr., J. M. Knutson, G. N. Mercer, and J. L. Detch, Appl. Phys. Lett., 4:180 (1964).
11. G. Convert, M. Armand, and P. Martinot-Lagarde, Compt. Rend., 258:4467 (1964).
12. W. E. Bell, Appl. Phys. Lett., 7:190 (1965).
13. G. R. Fowles and W. T. Silfvast, IEEE J. Quantum Electronics, QE-1:131 (1965); W. T. Silfvast, G. R. Fowles, and M. P. Hopkins, Appl. Phys. Lett., 8:318 (1966).
14. W. B. Bridges and A. N. Chester, IEEE J. Quantum Electronics, QE-1:66 (1965).
15. A. L. Bloom, Proc. IEEE, 54(10):1262 (1966).
16. W. R. Bennett, Jr., Appl. Optics, Suppl. Chemical Lasers, No. 3 (1965).
17. W. R. Bennett, Jr., P. J. Kindlman, G. N. Mercer, and J. Sunderland, Appl. Phys. Lett., 5:158 (1964).
18. J. Bakos, J. Strigetti, and L. Varga, Phys. Lett., 20:503 (1966).
19. H. N. Olsen, J. Quant. Spectrosc. Radiative Transfer, 3:59 (1963).
20. F. A. Horrigan, S. H. Koozekanani, and R. A. Paananen, Appl. Phys. Lett., 6:41 (1965).
21. H. Statz, F. A. Horrigan, S. H. Koozekanani, C. L. Tang, and G. F. Koster, J. Appl. Phys., 36:2278 (1965).
22. H. Statz, F. A. Horrigan, S. H. Koozekanani, C. L. Tang, and G. F. Koster, Proc. Conf. Physics of Quantum Electronics, Puerto Rico, McGraw-Hill, New York (1965), p. 674.
23. R. I. Rudko and C. L. Tang, Appl. Phys. Lett., 9:41 (1966).
24. H. Marantz, R. I. Rudko, and C. L. Tang, Appl. Phys. Lett., 9:409 (1966).
25. S. H. Koozekanani, IEEE J. Quantum Electronics, QE-2:770 (1966).
26. I. L. Bergman, L. A. Vainshtein, P. L. Rubin, and N. N. Sobolev, ZhÉTF Pis. Red., 6:919 (1967).
27. R. I. Rudko and C. L. Tang, J. Appl. Phys., 38:4731 (1967).
28. W. R. Bennett, Jr., G. N. Mercer, P. J. Kindlman, P. Wexler, and B. Hyman, Phys. Rev. Lett., 17:987 (1966).
29. M. Hammer and C. P. Wen, J. Chem. Phys., 46:1225 (1967).
30. R. K. Sundi, Sixth Conf. Ionized Phenomena in Gases, Paris (1963), p. 29.
31. S. G. Kulagin, V. M. Likhachev, M. S. Rabinovich, and V. M. Sutovskii, Zh. Prikl. Spektrosk., 5:534 (1966).

32. V. M. Likhachev, M. S. Rabinovich, and V. M. Sutovskii, ZhÉTF Pis. Red., 5:55 (1967).
33. E. T. Gerry, Appl. Phys. Lett., 7:6 (1965).
34. J. D. Shipman, Appl. Phys. Lett., 10:3 (1967).
35. J. D. Shipman and A. C. Kolb, IEEE J. Quantum Electronics, QE-2:298 (1966).
36. A. Javan, Phys. Rev. Lett., 3:87 (1959).
37. V. A. Fabrikant, Doctoral Dissertation, FIAN SSSR (1940).
38. J. H. Sanders, Phys. Rev. Lett., 3:86 (1959).
39. N. G. Basov and A. N. Oraevskii, Zh. Éksp. Teor. Fiz., 44:1742 (1963).
40. L. I. Gudzenko and L. A. Shelepin, Zh. Éksp. Teor. Fiz., 45:1445 (1963).
41. W. E. Lamb, Jr., and M. Skinner, Phys. Rev., 78:529 (1950).
42. W. R. Bennett, Jr., Ann. Phys., 18:367 (1962).
43. J. M. Hammer and C. P. Wen, Appl. Phys. Lett., 7:159 (1965).
44. E. I. Gordon, E. F. Labuda, and W. B. Bridges, Appl. Phys. Lett., 4:178 (1964).
45. S. Kobayashi, T. Izawa, K. Kawamura, and M. Kamiyama, IEEE J. Quantum Electronics, QE-2:699 (1966).
46. R. K. Leonov, E. D. Protsenko, and Yu. M. Sapunov, Opt. i Spektrosk., 21:243 (1966).
47. J. A. Bellisio, C. Fred, and H. A. Haus, Appl. Phys. Lett., 4:5 (1964).
48. L. J. Prescott and A. van der Ziel, Appl. Phys. Lett., 5:48 (1964).
49. W. H. Bennett, Phys. Rev., 45:890 (1934).
50. S. I. Braginskii and G. I. Budker, in: Plasma Physics and Controlled Thermonuclear Reactions [in Russian], Vol. 1 (1958), p. 186.
51. S. I. Braginskii and A. Migdal, in: Plasma Physics and Controlled Thermonuclear Reactions [in Russian], Vol. 2 (1958), p. 20.
52. S. M. Osovets, in: Plasma Physics and Controlled Thermonuclear Reactions [in Russian], Vol. 3 (1958), p. 165.
53. M. A. Leontovich and S. M. Osovets, Atomnaya Énergiya, 3:81 (1956).
54. L. A. Artsimovich, Controlled Thermonuclear Reactions [in Russian], Moscow (1961).
55. J. Hasted, Physics of Atom Collisions, Butterworths, London (1964).
56. G. Francis, Ionization Phenomena in Gases, Academic Press, New York (1960).
57. C. L. Longmire, Elementary Plasma Theory, Interscience, New York (1963).
58. M. N. Rosenbluth et al., Los Alamos Rep. LA-1850 (1954).
59. A. M. Andrianov, O. A. Bazilevskaya, and Yu. G. Prokhorov, in: Plasma Physics and Controlled Thermonuclear Reactions [in Russian], Vol. 2 (1958), p. 185.
60. S. I. Braginskii, I. M. Gel'fand, and R. P. Fedorenko, in: Plasma Physics and Controlled Thermonuclear Reactions [in Russian], Vol. 4 (1958), p. 201.
61. V. F. D'yachenko and V. S. Imshennik, in: Problems of Plasma Theory [in Russian], Vol. 5, Atomizdat (1967), p. 394.
62. J. E. Allen, in: Controlled Thermonuclear Reactions [Russian translation], Atomizdat (1960), p. 37.
63. J. D. Jukes, J. Fluid Mech., 3:275 (1957).
64. Yu. K. Zemtsov, V. D. Pis'mennyi, and I. M. Podgornyi, Dokl. Akad. Nauk SSSR, 155:312 (1964).
65. V. D. Pis'mennyi, Candidate's Dissertation, MGU, Moscow (1965).
66. A. M. Andrianov, O. A. Bazilevskaya, and Yu. G. Prokhorov, in: Plasma Physics and Controlled Thermonuclear Reactions [in Russian], Vol. 3 (1958), p. 182.
67. L. A. Artsimovich et al., Atomnaya Énergiya, 3:84 (1956).
68. V. M. Sutovskii, Candidate's Dissertation, FIAN SSSR, Moscow (1968).
69. G. G. Timofeeva and V. L. Granovskii, Zh. Éksp. Teor. Fiz., 30:477 (1956).
70. P. K. Cheo and H. G. Cooper, Appl. Phys. Lett., 6:177 (1965).
71. V. N. Kolesnikov, in: Proceedings of the P. N. Lebedev Physics Institute, Vol. 30: Physical Optics, Consultants Bureau, New York (1966), p. 53.

72. H. Griem, Plasma Spectroscopy, New York (1964).
73. V. I. Kogan, in: Plasma Physics and Controlled Thermonuclear Reactions [in Russian], Vol. 4 (1958), p. 258.
74. A. N. Vasil'eva, V. M. Likhachev, and V. M. Sutovskii, Zh. Tekh. Fiz., 39:341 (1969).
75. V. F. Kitaeva et al., Dokl. Akad. Nauk SSSR, 172:317 (1967).
76. V. N. Tsytovich, Zh. Éksp. Teor. Fiz., 51:1385 (1966).
77. D. E. T. F. Ashby et al., J. Appl. Phys., 36:29 (1965).
78. G. Sklizkov, Candidate's Dissertation, FIAN SSSR (1967).
79. W. H. McMahan and J. P. Bowen, IEEE J. Quantum Electronics, QE-2:567 (1966).
80. G. M. Malyshev, Zh. Tekh. Fiz., 35:2122 (1965).
81. F. Rostar, J. Phys., 27:367 (1966).
82. J. P. Dougherty and D. T. Farley, Proc. Roy. Soc., A259:79 (1960).
83. E. E. Salpeter, Phys. Rev., 120:1528 (1960).
84. J. A. Fejer, Canad. J. Phys., 39:716 (1961).
85. G. Herziger and S. Schmidt, Phys. Verhandl., 18(1):1 (1967).

PHYSICAL PROCESSES IN MOLECULAR HYDROGEN, DEUTERIUM, AND FIRST-POSITIVE-BAND-SYSTEM NITROGEN PULSED GAS-DISCHARGE LASERS*

I. N. Knyazev

INTRODUCTION

The growth of quantum radiophysics has placed considerable priority on the problem of finding active substances and investigating the physical processes that evolve in those substances. The efforts aimed at finding new active media have expanded the number of transitions at which lasing is possible and have disclosed new opportunities for the transformation of excitation energy into energy of stimulated laser emission. The investigation of the physical processes involved has, in its own right, the ultimate goal of answering important questions concerning the fundamental parameters of the active medium, as well as the capabilities of particular techniques for the inducement of population inversion. Investigations of this sort assume special significance in application to certain types of lasers. Among the latter are pulsed gas-discharge lasers ("gas lasers"), whose active medium is the sparsely investigated low-temperature plasma produced by a high-voltage pulsed discharge in gases.

The earliest experiments in the pulsed excitation of gas lasers were performed on a neon–helium continuous laser, from which emission was obtained in [1]. Even at relatively low excitation powers the significant advantages of the pulse method were noticeable, namely the comparative ease of lasing and a measurable increase in the peak power [2]. These advantages were more fully appreciated with the use of high-voltage pulsed excitation [3-5].

It is important to note that pulsed excitation was used in one form or another to obtain population inversion even before the advent of lasers. Typical examples are the "pulsed inversion" [6] and "fast adiabatic transmission" [7] techniques used in the microwave (maser) range. In both cases the upper working levels were filled by means of a superhigh-frequency emission source. Another technique used for the pulsed excitation of inversion in the optical range entails heating or cooling of the system in thermodynamic equilibrium [8]. The use of a nonequilibrium pulsed-discharge plasma for this purpose is still a highly efficient, and in fact the only, method for realizing inversion.

---

* Abridged text of the author's dissertation presented September 16, 1968, toward fulfillment of the academic degree of Candidate of Physicomathematical Sciences under the direction of the late Prof. P. A. Bazhulin and G. G. Petrash.

The use of high-voltage pulsed excitation led to the discovery of stimulated emission from dozens of new active substances and hundreds of transitions, as well as to a sharp increase in the peak radiated power and considerable broadening of the spectral interval in which gas-discharge lasing is possible. Included among the many pulsed gas lasers is an extensive group in which lasing originates at the leading edge of the current pulse, i.e., at the very beginning of filling of the working levels. Of particular interest are pulsed lasers of this type operating on electron transitions in diatomic molecules, which account for the most powerful pulsed gas lasers known today. Inherent in these lasers are certain common singularities of inversion excitation, which are associated with the specific populating of the molecular levels in a pulsed discharge.

At the time we began our own investigations lasing had already been observed in the diatomic molecules N_2 [3, 4] and CO [9]. The problem of finding new active media and laser emission lines was particularly urgent at that time. As a result of our subsequent research we were the first to discover pulsed lasing at electron transitions in H_2, D_2, and HD molecules and at approximately 90 new emission lines of the first positive (1+) band system of N_2 [5, 10, 11]. The investigation of these lasers constitutes the topic of the present article.

In the immediate wake of our first exploratory studies we were confronted with the problem of formulating a comprehensive description of the newly discovered phenomena before we could hope to obtain a quantitative interpretation of the experimental results. The exceedingly scant data available in the literature failed to provide the requisite information on the physical processes in the lasers. In the meantime, we had to solve important problems relating to the inversion mechanism, the laser-level excitation dynamics, and the power aspects of pulsed gas lasers. It was necessary to elucidate the problem of the ultimately attainable laser parameters, the most important of which are the efficiency, power, and emission pulse energy, not to mention the attainable population inversion.

Due to the complexity and relative ignorance concerning the physical processes involved in pulsed discharge it was virtually impossible to launch immediately into any sort of analysis at the quantitative level. The difficulties associated with quantitative analysis were chiefly attributable to the inadequate investigation at that time of the filling and de-excitation of the molecular states under powerful pulsed discharge conditions. It was unknown to what extent step-by-step filling and de-excitation of the laser states could act. Nor were the time dependences of the electron density and effective electron temperature understood, so that it was difficult to analyze the corresponding equations for filling of the laser states. The answer to the question of the inversion mechanism in lasers operating in the 1+ band system of N_2 was unclear.

The comprehensive interpretation of the effects involved in pulsed gas lasers poses an extremely acute problem at the present time. On account of the diversity of physical processes taking place in these lasers it would be unrealistic to hope for an immediate explanation of all the lasing cases known to date. These considerations impose a special timeliness on the detailed investigation of individual lasers or of a particular category of lasers. The present study is viewed as a step toward the solution of this general problem.

CHAPTER I

STATE OF THE ART OF RESEARCH ON PULSED GAS-DISCHARGE LASERS; STATEMENT OF THE PROBLEM

§1. General Description of Pulsed Gas Laser Research

Although not much time has elapsed since the first experiments on the pulsed excitation of gas lasers, the total volume of published data relating to the problem is rather extensive. This situation is attributable primarily to purely quantitative factors, namely the large number of active substances in which lasing has been detected and the tremendous number of observable emission lines. Each individual case of lasing has actually been studied relatively little. All of these considerations render it exceedingly difficult to make an exhaustive survey of the literature as is usually done in such cases. We shall, instead, seek to give a general description of research on pulsed gas lasers, with the intention, on the one hand, of demonstrating the status of the present investigations in the overall program relative to pulsed gas lasers and, on the other hand, of focusing the maximum attention on those problems which have direct bearing on the topic of the present study.

Pulsed excitation was used in early experiments on the generation of continuous emission from a neon–helium laser [1, 12]. In one paper an increase in gain was even observed upon cessation of discharge [1]. No real significance, however, was attached to the observed effects. The interest in pulsed excitation mounted somewhat after lasing had been obtained at the leading edge and in the afterglow of a low-current pulsed discharge in a neon–helium laser [13]. Our own more detailed investigation [2] showed, in particular, that lasing at the leading edge of the current pulse is associated with direct excitation of the neon working levels by electrons from the ground state during the incipient development of pulsed discharge ("direct electron impact" hypothesis). It was noted even under the conditions of low-power excitation that pulsed lasing is observed over a considerably broader range of experimental conditions and at more transitions, and has a somewhat higher power than in continuous excitation.

However, the principal interest in pulsed excitation stemmed from the use of a powerful pulsed discharge in gas lasers. The use of high-voltage pulsed excitation made it possible to raise the radiated power of a neon–helium laser by about three orders of magnitude, to approximately 100 W [14]. Soon afterwards pulsed lasing was discovered in new gases: molecular nitrogen (at infrared transitions of the first positive (1+) band system [3] and at ultraviolet transitions of the second positive (2+) band system [4]), CO [9], Ar [15, 16], and certain other gases. At about the same time lasing was also discovered in H_2, D_2, and HD [5, 10], as well as at several lines in the visible part of the neon spectrum [17] and at a set of new transitions of the 1+ band system of N_2 [11, 18]. An unprecedented peak power for that time, about 250 W, was obtained in molecular-hydrogen gas lasers.

After the first studies on pulsed gas lasers there was a virtual onslaught of more and more new lasers of this type using the most diverse working media over an extraordinarily broad spectral region. Some idea of the scope of these new discoveries may be gained from the surveys [19, 20], [21] (spectral range from 15 to 337 μ), [22] (ionic lasers), and [23] (diatomic molecules).

The current published data suggest a whole series of advantages for the high-voltage pulsed excitation of gas lasers.

The most important thing to consider is that the pulsed excitation method afforded an extremely powerful and to a certain extent universal technique for the generation of inversion in a gas discharge. Unlike continuous gas lasers, in which preeminently inert gases are used, pulsed gas lasers use the widest imaginable variety of substances. Pulsed lasing is frequently observed from several ions of the same atom. Pairs of metal or gaseous compounds containing a required element serve as auspicious active media. Pulsed lasing has now been observed from about 30% of all elements of the periodic table. No other excitation technique has yielded such eminent results.

Lasing in the submillimeter (~ 0.4 mm) [21] and ultraviolet (~ 2300 Å) [22, 24] regions of the spectrum was first obtained specifically by the pulsed excitation of gas lasers. The whole span of electromagnetic waves from the submillimeter to the ultraviolet is densely filled with a large number of observed emission lines of pulsed gas lasers.

While retaining the homogeneity of the active medium inherent in continuous gas lasers, pulsed gas lasers have a far higher power level, comparable with the power level of solid-state lasers. The maximum peak power of pulsed gas lasers is continuously growing, having now attained a value of 2.5 MW [25]. Given the same peak power, the energy spent in the excitation of a pulsed gas laser in the free-oscillation regime is two or three orders lower than in the case of solid-state lasers using optical pumping. The threshold energy of pulsed gas lasers is also considerably lower than for solid-state lasers, amounting to, in general, tenths or even hundredths of a joule.

The situation for pulsed gas lasers is more favorable with respect to efficiency than for many continuous gas lasers. As is known [26], the latter type of laser, with rare exception, has 0.1% efficiency. Two conditions are necessary in order to obtain higher efficiencies in pulsed excitation, at least in excitation by the "direct electron impact" technique [27]. First, it is required to use relatively low-lying atomic-molecular states, which have the greatest effective cross section for excitation by electrons [28]. Second, it is required that the ratio of the energy quantum at the working transition to the excitation energy for the upper laser state be close to unity. In this case, clearly, it should be possible to raise the efficiency of pulsed gas lasers to several percent [29]. To realize this efficiency in continuous lasing requires the fulfillment of more stringent demands [30, 31].

The repetition rate of the emission pulses of pulsed gas lasers can attain 13 kHz with streaming through the discharge tube [32]. In optically driven solid-state lasers, as we know, considerable difficulties are involved in trying to increase the repetition rate. The ease with which pulsed gas lasers can be made to operate at fast repetitions is particularly valuable insofar as this is the kind of operation required for certain problems of a practical nature. On account of the increased repetition rate it is entirely possible to obtain a high average radiated power from pulsed gas lasers.

It is important to note, in addition, the exceptional simplicity of pulsed gas laser devices and their operational reliability. Abundantly available and inexpensive gases, such as nitrogen and hydrogen for example, can be used for the active medium, thus casting a favorable light on pulsed gas lasers by comparison with semiconductor and certain other types of solid-state lasers.

The foregoing properties of pulsed gas lasers invest this type of laser with considerable promise from the standpoint of potential practical applications. Pulsed gas lasers have broad applicability for the solution of problems relating to optical detection, communications, and holography.

Despite its unique attributes, the pulsed excitation of inversion in a gaseous discharge has so far been studied only meagerly. The investigation of pulsed gas lasers is presently still very much in the nascent phase. With infrequent exception research is aimed primarily at finding new transitions and active media. Experience has shown that to accomplish this does not require incisive preliminary studies. The purely experimental method is the most effective and expedient means of uncovering new active substances. This expediency has naturally diverted the attention of researchers from the physical processes involved in pulsed gas lasers, particularly up against the difficulty of the requisite investigative work under pulsed discharge conditions.

As a rule, qualitative hypotheses are invoked to account for the laser action. In many cases the lasing mechanism remains unknown, even qualitatively. It is not clearly understood whether there is any possibility of improving the experimentally achieved parameters of pulsed gas lasers or whether the power, population inversion, and efficiency are the best we can expect from this type of laser. The investigation of the physical processes underlying pulsed gas lasers is becoming an increasing urgent problem. The continued development of these lasers and the elucidation of the ultimate capabilities of the pulsed excitation method for population inversion in gaseous discharges is inconceivable without the solution of this problem.

Of the many pulsed gas lasers known at the present time we have selected for detailed investigation the cognate class of lasers operating on molecular hydrogen and the isotopic D_2 and HD molecules, as well as $N_2(1+)$ lasers, i.e., lasers acting in the first positive band system of molecular nitrogen. This choice was dictated in part by the logical desire to probe more fully into the effects disclosed by our exploratory studies. Moreover, it was clear that the most complete information about the physical processes in a pulsed discharge could be obtained specifically from molecular systems having a rich stimulated and spontaneous emission spectrum. Of particular interest in this sense is lasing from isotopic molecules, the use of which promotes not only a more precise allocation of the emission lines to given molecular transitions [33], but also the acquisition of valuable additional information concerning the excitation processes. In the case of $N_2(1+)$ lasers we were intrigued by the tremendously large effective cross section for excitation of the upper working states of these lasers: about 10^{-16} cm^2 [34, 35]. If this type of laser is excited by electrons, the most general considerations indicate the possibility of an enormous peak power output. Even more important in terms of efficiency is the ratio of the emission energy quantum to the energy of the upper laser level. For the 1+ band system of N_2 this ratio (~ 0.2) is only about half that of the CO_2 laser. We point out for comparison that the indicated ratio is equal to ~ 0.05 for the neon–helium laser.

§2. Brief Survey of the Literature on Lasers Operating on Electron Transitions in Diatomic Molecules; Statement of the Problem

At the present time there are nine known lasers of approximately the same type as lasers acting in H_2 and the 1+ band system of N_2. The stimulated emission of all these lasers is observed at transitions between electron states of diatomic molecules. Hereinafter we make special use of the published data concerning lasers operating on the 1+ and 2+ band systems of N_2 [$N_2(1+)$ and $N_2(2+)$ lasers]. More detailed information on lasers operating on electron transitions in diatomic molecules is given in [23].

Pulsed laser action in H_2, D_2, and HD molecules has been investigated in [5, 10, 36]. Lasing in these molecules was also observed in [37, 38], somewhat later than our work.

Lasing in molecular nitrogen was first observed in [3] in the near infrared (0.75 to 1.2 μ) at about thirty lines simultaneously in the 1+ band system of N_2. The accuracy of the wavelength measurements was low (about 0.4 Å). The peak radiated power attained 100 W. A stimulated emission pulse ~ 1 μsec wide was observed at the end of the excitation pulse. The properties of the emission were not studied in detail. The hypothesis of step-by-step filling of the upper laser state in intermolecular collisions was used in [3] and later in Bennett's survey [19] to explain the emission. The same hypothesis was proposed by the authors of [39] quite some time ago (in the thirties) for the interpretation of certain observed phenomena.

Another point of view with regard to the inversion mechanism in the 1+ band system of N_2 was offered in [40]. Proceeding from indirect considerations, the authors of this paper deduced that the working levels of an $N_2(1+)$ laser were filled by "direct electron impact" from the ground state (independently of [40] we used the "direct electron impact" mechanism to explain H_2 lasing [5]). The investigation of the laser characteristics for narrow tubes using a relatively low driving current showed that with an increase in the voltage the radiated power at first increases, then begins to decrease. An increase in the pressure brings about a certain increase in the radiated power. The experimental conditions in [40] were not varied too broadly.

A fairly recent paper [41] was devoted chiefly to a careful investigation of the emission spectrum of an $N_2(1+)$ laser. The spectrum was investigated in this case using a highly perfected spectrograph at the National Bureau of Standards (USA), which has an excellent reputation for superior technical achievement in work of this kind. The principal results of this study with regard to the rotational structure of the emission spectrum almost exactly coincide with the results of our work [11] published a year and a half earlier. Interesting information is given in [41] on the operation of lasers in the double-pulse excitation regime, permitting an assessment of the attainable repetition frequency. The inversion mechanism is not discussed in [41].

The three papers cited above exhaust the published data on $N_2(1+)$ lasers at the time the present article was written. Obviously, the data are very scant. The stimulated emission lines had not even been identified. In general, there had not been any interpretation of the rotational structure of the emission spectrum, and the allocation to vibrational transitions could only be regarded as conjectural, whereas for the reliable allocation to molecular transitions an accuracy of ~ 0.4 Å was unquestionably deficient. In light of the absence of a reliable interpretation of the emission spectrum the diagram of the laser levels remained equally unclear. But without the knowledge of this diagram it would be impossible to undertake any serious discussion of the problem associated with the inversion mechanism. The laser characteristics had not been adequately studied. It was also uncertain whether a large radiated power could be obtained or how the variation of particular experimental conditions, say, the diameter and length of the discharge tube, would affect the emission intensity.

There was also complete uncertainty in the very important problem of the inversion mechanism in the $N_2(1+)$ laser. About all that existed in this respect, essentially, were extremely coarse qualitative postulates, which were mutually conflicting. The pulsed nature of the emission was accounted for differently in [3] and [40]. It was assumed in the former that emission cutoff takes place due to an increase in the population of the lower laser state as a result of spontaneous and induced radiative transitions from the upper laser levels. In [40], on the other hand, the pulsed nature of the emission was attributed to the fact that the direct excitation of the upper laser state is realized only for a short time interval at the instant of breakdown of the gas in the discharge tube, while the plasma is still made up of fast electrons.

The situation was somewhat improved in this respect for the case of the $N_2(2+)$ laser ($\lambda = 3371$ Å), from which lasing was first detected in [4]. A higher voltage was required for stimulated emission at the ultraviolet transitions of N_2. The development of the pulsed discharge in this case occurs far more rapidly than in the case of the $N_2(1+)$ laser, so that the physical conditions in these two lasers are significantly dissimilar. The initial very small radiated power (10 W) [4] was elevated by the perpendicular-discharge, or "crossed-field" technique to 200 kW with a pulse energy of 4 mJ [42]. The emission spectrum of the $N_2(2+)$ system was investigated in [43] and, somewhat later, in [41]. A perfected "crossed-field" technique was used to increase the peak power of this laser to 2.5 MW [25] (at a pulse energy of 7 mJ). Proceeding from the assumption of direct excitation of the working levels by ground-state electrons, Gerry [44] succeeded in performing a quantitative calculation of the power density of an $N_2(2+)$ laser at saturation as a function of the time and in obtaining satisfactory agreement with experiment. The approximations used in this study and, more particularly, the results of calculations carried out by numerical methods for basically one set of experimental conditions, generally speaking, cannot be used for other pulsed gas lasers or, specifically, for $N_2(1+)$ lasers. Of all the processes of de-excitation of the upper laser state, only radiative decay was taken into account, for example, in [44]. For an $N_2(1+)$ laser this is clearly invalid, because the radiative lifetime of the upper working levels of this laser (~ 10 μsec) exceeds the stimulated emission period (0.2 to 1.5 μsec) by more than one order of magnitude. It was also assumed in the calculations of [44] that the increase of the current with time is completely determined by the inductance of the discharge circuit. Inasmuch as the development of discharge in the $N_2(1+)$ laser proceeds far more slowly, the inductance, on the contrary, can have only a slight influence on the behavior of the current growth.

Of other pulsed lasers operating on electron transitions in diatomic molecules, the CO laser has been more thoroughly investigated [9, 23, 40, 45-47]. Relatively low-power lasing (8 W) is observed for this molecule in the visible part of the spectrum (6500 to 4500 Å). We note also the lasing at molecular transitions of N_2 in the spectral interval from 3 to 7 μ [48].

Using certain singular features of the spectral composition and time development of the lasing process, we first conducted a qualitative analysis of the processes in H_2, D_2, HD, and $N_2(1+)$ lasers. It turned out that the principal experimental data are satisfactorily explained on the basis of the mechanism of excitation of the laser levels up from the ground state by "direct electron impact." After these preliminary investigations we had a clearer picture of the difficulties ahead with regard to quantitative analysis.

In listing the foremost of these difficulties we stress the fact that despite the auspicious interpretation of the characteristics of the spectrum and time development of the stimulated emission pulse, the inversion excitation mechanism in the $N_2(1+)$ laser was still speculative. The main reason for this enigma was that all arguments for or against a particular inversion mechanism were of a purely qualitative nature.

Furthermore, it was not known exactly how the population of the laser levels grows with time. It is essential to note that the processes of excitation of atomic-molecular states under relatively low-temperature conditions (5 to 10 eV) in a strongly nonequilibrium and nonstationary plasma, as is the case with the pulsed-discharge plasma in a pulsed gas laser, have been almost totally ignored to date. These processes did not seem particularly significant either with respect to the physics of breakdown, the central consideration of which has always been problems relating by and large to the breakdown potential [49], or with respect to controlled thermonuclear reaction studies, the principal object of which is the already fully ionized plasma;* both of these aspects of plasma physics are related in one way or another to a pulsed discharge similar to that used in pulsed gas lasers.

As the population of atomic-molecular states is increased in a powerful pulsed discharge step-by-step processes become possible in the filling and de-excitation of the laser levels, and it is exceedingly difficult to account for these processes in general form. In order to analyze the operation of the laser it is important to know exactly what processes actually lead to the cutoff of inversion under pulsed-discharge conditions.

---

* Lasing in ionized atoms has recently been produced in a pinch discharge [50]. The results of an investigation of lasers of this type are presented in the preceding article of this collection.

In connection with the fact that lasing is initiated at the leading edge of the current pulse, i.e., at the instant of breakdown of the discharge gap, the role of breakdown effects in the inversion formation process was still not clear. The situation was compounded further by the fact that, as it developed, the transient dynamics of the pulsed discharge in pulsed gas lasers had not been sufficiently investigated. This engendered still another problem. Due to the uncertainty involved in gas-discharge phenomena the time dependence of the electron density (n_e) and effective electron temperature (T_e) were unknown, and without knowledge of them any analysis of the corresponding population equations was plainly out of the question. For a homogeneous field the relationship between the current density and electron density is very simple. However, on account of the lack of information on the nature of the current growth and potential distribution over the length of the tube at the instant of lasing it proved impossible to draw any conclusions regarding the form of the function $n_e(t)$. Under real conditions the nature of the current growth and the form of $n_e(t)$ can be determined by cathodic processes. All of this renders impossible the *a priori* selection of a physical model that could be used to describe the pulsed discharge in calculations.

The information needed for calculations with regard to the transition probabilities and effective cross sections for electron excitation of the levels is generally unavailable. Comparison with experiment is further complicated by the fact that the calculation of the power and energy of the emission pulse, which are customarily measured in experiments, is inherently complex, and there are no data on the gain and efficiency of inversion excitation. It is certainly not surprising, in this light, that a quantitative analysis has so far only been carried out for one case, namely for the $N_2(2+)$ laser [44].

In view of the lack of requisite data on the probability of radiative transfer between laser levels and the Franck–Condon loci for the entire series of electron-vibrational transitions, it was impossible to conduct a quantitative analysis of the operation of H_2, D_2, or HD lasers. Only for $N_2(1+)$ lasers was a reasonably complete set of necessary data available.

All of the foregoing considerations lucidly spell out the problem area of the present study. In the initial phase of our investigations we solved the problem of finding new active media and emission lines and of optimizing the excitation conditions. Also part of the problem was to explain the qualitative inversion mechanism in $N_2(1+)$ lasers and in H_2, D_2, and HD lasers. In this phase we were guided by the following objectives:

(1) to develop a high-voltage excitation system and concomitant detection-recording devices;
(2) to investigate the characteristics of the lasers;
(3) to measure the wavelengths of the stimulated emission lines with the accuracy demanded for reliable allocation to vibrational-rotational transitions, and to elucidate the laser-level diagram.

The remainder of the program must be directed primarily toward $N_2(1+)$ lasers:

(1) to investigate quantitatively the excitation of the working laser levels, and to assess the efficiency of possible channels of radiative filling;
(2) to study the de-excitation of the upper working laser state and the resulting cutoff of inversion under real pulsed-discharge conditions;
(3) to explicate the role of breakdown phenomena in the formation of inversion, investigate the physical conditions in the discharge at the instant of lasing, and formulate and analyze a physical model of the pulsed-discharge effect;
(4) to investigate the current buildup process;
(5) to measure the gain at individual rotational lines with time scanning;
(6) to perform a quantitative analysis of the operation of the $N_2(1+)$ laser, and to calculate the attainable laser parameters, including the population inversion, efficiency, peak power, and energy of the laser pulse.

CHAPTER II

EXPERIMENTAL METHODOLOGICAL PROBLEMS

§1. The Laser and Its Excitation Circuit

The laser was driven by periodic discharge of a capacitor through a discharge tube in conjunction with a three-electrode air gap for current switching. Unlike driver systems using thyratrons and pulse transformers [3, 4, 51], this circuit makes it possible to obtain higher currents, because current of virtually any magnitude can be sent across the spark gap, whereas the current through a thyratron is limited. For the present investigations, naturally, we were motivated to use whatever driver system would admit the maximum range of variation of the discharge parameters.

The laser excitation circuit is shown in Fig. 1a. A capacitor rated from 0.01 to 0.15 μF is charged by a VS-50-50 rectifier through the resistance $R_3 \sim 1$ MΩ to a voltage regulatable from 1 or 2 to 50 kV. We used type IM 110-0.01 capacitors with an inductance of about 2 μH, as well as low-inductance capacitors with an internal inductance of 30 nH. The maximum energy stored by the working capacitor did not exceed 40 J. The capacitor charging time $\tau = C \cdot R_3$ determined the current pulse repetition rate. With a 0.01-μF capacitance the repetition rate sometimes attained 20 Hz. The maximum working voltage was limited by the capabilities of the rectifier.

The firing pulse is sent to the three-electrode spark gap by discharging the capacitor across the laser tube. An air gap with brass main electrodes was used. The separation of the electrodes was chosen as a function of the working voltage and could be extended to 20 mm. The firing pulse was fed to the third electrode, which was encased in a hole drilled coaxially into one of the main electrodes. The firing system was assembled from thyratrons and was controlled from a GIS-2 or GI-1 pulse generator. The latter generator was capable of operating in the one-shot triggering mode. Experience showed that a standard three-electrode spark gap is sufficiently stable in switching of the discharge loop at a frequency of 20 Hz for

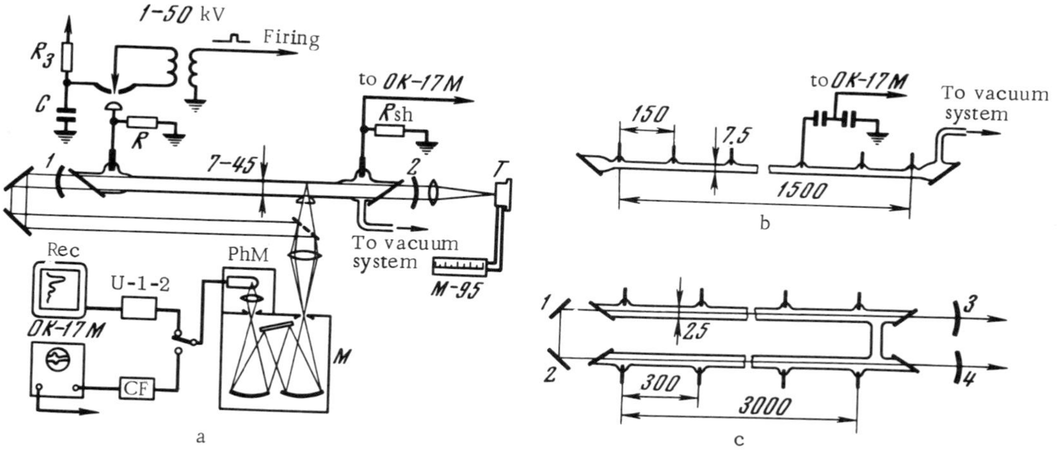

Fig. 1. Schematic of the experimental arrangement. Fig. 1a: T) thermopile; M-95) mirror galvanometer; CF) cathode follower; M) monochromator; Rec) recorder; U-1-2) dc amplifier; PhM) photomultiplier; OK-17M) oscilloscope; R = 30 kΩ; R_{sh} = 0.01 Ω. For the investigation of the laser characteristics the dielectric mirrors 1 and 2 have respective transmissivities of 1 and 17%.

a pulse energy of ~ 10 J. Measurements showed that the internal impedance of the gap is approximately two orders of magnitude smaller than the resistance of the discharge tubes, so that almost all the energy of the capacitor is transmitted into the discharge tube.

The experiments were carried out with a gas laser using external mirrors. Quartz windows were bonded to the discharge tube at the Brewster angle by means of picein or shellac. We used mainly dielectric mirrors, whose transmissivity and range of maximum reflectivity could be widely varied to suit the problem. We also used aluminum-coated or silvered mirrors. The transmissivity of the mirrors was measured on a Hitachi spectrophotometer. The losses of the cavity with dielectric mirrors were determined mainly by the transmissivity of the mirrors, and the absorption losses did not exceed 2 or 3%. As for the cavities using aluminum- or silver-coated mirrors, the absorption losses could be considerable, particularly with the insertion of intermediate mirrors. For example, the losses of the cavity illustrated in Fig. 1c with four aluminum mirrors exceeded 80%.

The laser characteristics were measured for the most part in discharge tubes with an effective length of 1.3 m (see Fig. 1a) made of 3S-5 glass. The diameters of these tubes ranged from 7 to 45 mm. We used a confocal cavity with dielectric mirrors 2 m in radius.

We used longer tubes of quartz with a special construction (Figs. 1b and 1c) to explicate the influence of the discharge geometry on the lasing power and to investigate the breakdown processes. The length of the discharge gap of these tubes could be varied over wide limits by switching of the electrodes.

We used internal electrodes of thoriated tungsten or aluminum, which were soldered into side branches of the tube. All the discharge tubes were mounted on a grounded metal baseplate.

Special measures were not taken to stabilize the temperature of the discharge tubes. The mean temperature of the tube walls at the most commonly used repetition rates generally exceeded room temperature only by 10 to 20°C.

The working tubes were evacuated by an ordinary glass vacuum system incorporating an oil-vapor pump to pressures of 10^{-5} or 10^{-6} mm Hg and were then filled with high-purity nitrogen. Hydrogen was admitted to the tube by means of a palladium injection device. The pressure was measured in the 0.2 to 10 mm Hg range by an oil manometer and in the range below 0.2 mm Hg by a specially calibrated thermocouple manometer. The pressure measurement error was about 10% in the most important interval around 0.7 mm Hg.

The purity of the gas was monitored with an STÉ-1 spectrograph. The resulting spectrograms were also used for qualitative analysis of the pulsed-discharge plasma and a determination of the strongest system of bands in the spontaneous emission from molecular nitrogen in the interval from 2300 to 9000 Å.

The average radiated power was measured with a calibrated thermopile. The peak power was calculated from the measured average power, emission pulse width, and pulse repetition rate. To protect the thermopile against blackening due to hole-burning at peak lasing power we used neutral filters having a known transmissivity.

For time scanning of the stimulated or spontaneous emission pulse a light pulse was transmitted through a monochromator to a photomultiplier and then through a cathode follower to one beam of an OK-17M double-beam oscilloscope. The current pulse was normally transmitted to the second beam (see Fig. 1a).

The detector was usually an FÉU-22 or FÉU-17 photomultiplier for the infrared and near ultraviolet regions of the spectrum, respectively. The time resolution of the detection-recording system was limited by the photomultiplier and varied from 50 to 100 nsec for different photomultipliers. Time-integrated recording could be realized by connecting an integrating circuit, U-1-2 dc amplifier, and recording potentiometer to the photomultiplier output.

Careful attention was devoted to undistorted signal recording during the measurements. The relatively small interval of linear deflection of the OK-17M oscilloscope beam necessitated special precautionary measures. The light signal was attenuated by neutral filters, and the current or voltage pulse by a calibrated resistance voltage divider. For pulse width measurements a sine wave was transmitted from a GSS-6A oscillator to one of the OK-17M beams. An analysis of the oscillograms showed that a large part of the sweep was linear and identical for both beams. The pulse from a GIS-2 generator was fed directly to both beams to test the synchronization of the beams.

§2. Spectral Measurements

The investigation of molecular gas spectra, which are normally marked by considerable complexity, calls for spectral instrumentation of superior quality. Of the lasers investigated, molecular nitrogen has a particularly complex spectrum.

The emission spectra in our experiments involving precise wavelength measurements were recorded photographically on a DFS-3 instrument with a 1200-line/mm grating and focal length of 4 m. The emission was observed in the first order of the grating, where the dispersion was 1.7 Å/mm. Infrarot 750, 850, and 950 photographic plates, which are sensitive in the near-infrared region of the spectrum, were used to record the spectra. A singular feature of the measurements was the fact that the distance between the light source (laser) and spectrograph was very large, about 50 m. For this reason we gave careful attention to the adjustment of the optical system. The random shifts that can easily arise in this case were eliminated by measuring the emission spectrum on a large number of plates. For the reference spectrum we used the infrared spectrum of neon, which has relatively well-spaced infrequent lines (the wavelengths of the neon lines are known from [52]). The latter consideration somewhat increased the experimental error, which nevertheless did not exceed 0.09 cm^{-1}. Inasmuch as the spontaneous emission lines of hydrogen and deuterium, judging from the tables of [53], are generally spaced at a distance of 1Å, all of the recorded stimulated emission lines of these molecules could be referred with certainty to spontaneous emission transitions with reliably measured (in [54, 55]) wavelengths. A large part of the stimulated emission lines of molecular nitrogen could also be identified with corresponding spontaneous emission transitions in N_2, for which the wavelength data are known from [56, 57] to within a few hundred angstroms. Some of the identified stimulated emission lines of N_2 with wavelengths known from [56, 57] were then used as reference lines, which were more densely spaced than the neon lines. This procedure somewhat increased the measurement accuracy, making it possible to interpret the rest of the stimulated emission lines.

For coarse measurements of the wavelengths of the stimulated emission lines correct to 5 or 10Å we also used a monochromator fabricated in the laboratory. The instrument used a Fastie circuit and had a 300-line/mm grating, focal length of 650 mm, and relative aperture of 1:4.5. The monochromator was used extensively for measurements of the absolute and relative populations of the laser states, for measurements of the relative intensity of the emission lines, for identification of individual rotational lines in gain determinations, and for a determination of the parts of bands in spectra already identified on other instruments.

Essentially two variables are very important in all of these problems, namely the spectral interval and instrument function of the monochromator. If we neglect mirror aberrations, the calculated resolvable spectral interval at the first grating order is 0.4Å. Under actual conditions, due to the unsatisfactory quality of the mirrors and the ensuing aberrations, the resolution of the monochromator is considerably worse. It was mandatory to strongly diaphragm the mirrors in order to reduce the aberrations. Under these conditions we were able to attain a resolution of 0.5 Å at the third grating order. Concerning the instrument function, for a slit width of 100 to 300 μ, as was normally used in the measurements, this function was nearly triangular. A variety of filters, including the interference type, was used to separate the orders. The transmissivity of the filters was measured on a Hitachi spectrophotometer. The measurements in the infrared region were performed at the second and third grating orders and in the ultraviolet at the fifth order. For measurements of the relative population of the vibrational levels of the N_2(1+) band system and the relative population of the levels of the 1+ and certain other band systems of nitrogen (with primary concern for the 2+ band system of N_2) it is important to know the contours of the bands and the sensitivity of the recording system. Two methods were used to measure the band contours: measurement of the time-integral band contour and time scanning. In the former case an integrating network was connected to the photomultiplier output, and the signal was then fed through a dc amplifier or balanced cathode follower to a recording potentiometer (see Fig. 1a). In the latter method the monochromator was used to segregate individual parts of the band, and the signal was recorded on an oscilloscope. The time variation of the band contour was easily calculated from the resulting set of oscillograms. In either case the instrument function of the monochromator was made considerably smaller than the half-bandwidth.

A calibrated band lamp with a known brightness temperature was used to determine the sensitivity of the recording system. The energy flux per unit solid angle from 1 cm^2 of surface of the band lamp per second in a unit wavelength interval is as follows (brightness $B_{\lambda,T}$):

$$B_{\lambda,T} = \varepsilon_\lambda \frac{2c^2}{\lambda^5} \frac{h}{e^{hc/kT\lambda} - 1}, \tag{1}$$

where ε_χ is a correction for the actual emittance of the tungsten strip [58]. The relative sensitivity of the recording system is readily determined from the known brightness $B_{\lambda,T}$. The relative sensitivity thus determined was checked by determining the relative intensity of certain bands for which these relative intensities are known from the Franck–Condon loci.

The absolute intensity of spontaneous emission at vibrational transitions was measured by a comparative method (using the measurement circuit in Fig. 2). The light from the band lamp was modulated and recorded oscillographically. The absolute intensity of spontaneous

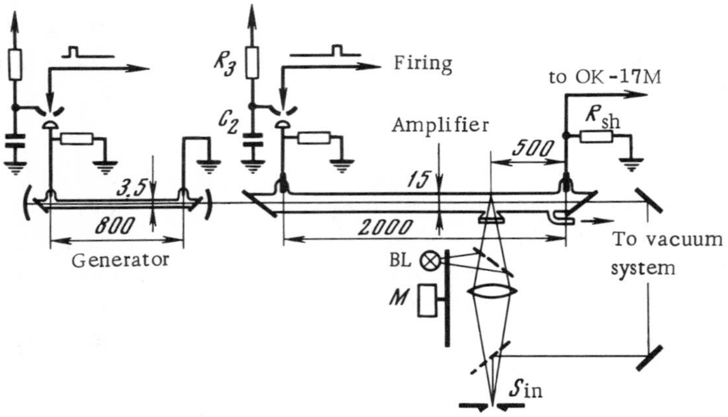

Fig. 2. Schematic of system for measuring absolute population of the working levels and gain of a laser. $C_1 = 0.01\,\mu F$; $C_2 = 0.12$ to $0.08\,\mu F$; S_{in}) entrance slit of monochromator; BL) band lamp; M) modulator.

emission from the volume V of the discharge tube at a selected band is easily calculated from the known brightness of the band lamp, the known wavelength interval $\Delta\lambda$, which is determined by the slit width, from the premeasured band contour, and from the relative intensity of the modulated light from the band lamp and of the spontaneous emission in the interval $\Delta\lambda$. The measurement geometry is such that the solid angle in which light is emitted is approximately the same for the band lamp and discharge tube. Only the parts of the plasma volume whose emitted radiation reaches the detector are included in the calculation of the volume V.

Knowing the time-varying absolute intensity of the integral emission in the selected band and the volume of the plasma from which that emission emanates, one can easily calculate the absolute population of the corresponding vibrational state as a function of the time. The error in the determination of this population is limited principally by the measurement accuracy of the absolute intensity and is estimated to be about 50%.

The gain at individual rotational lines was measured with time scanning by the circuit shown in Fig. 2. The generating source is a laser with a tube 4 mm in diameter, an active length of 80 cm, and mirrors having a transmissivity of ~ 4% and a radius of curvature of ~ 15 m. The small diameter of the discharge tube and large radius of the mirrors ensured light production in a relatively narrow and weakly divergent beam. The spectrum of the generator under these conditions turned out to be greatly simplified, lasing occurring primarily at the most intense lines. The lasing spectrum was recorded on an STÉ-1 spectrograph. The individual rotational lines were easily resolved with the monochromator. The beam emanating from the generator was directed along the center of the amplifier tube and then to the monochromator, which was used to separate out the rotational line. The whole system was tuned by means of a neon–helium laser at $\lambda = 0.63\,\mu$. The lasing pulse was shifted relative to the current pulse in the amplifier tube by means of a GIS-2 pulse generator, so that the gain (absorption) was measured as a function of the time. The time resolution in this case was actually determined by the width of the generator pulse and was approximately 0.2 μsec. The synchronization stability of the generator pulse relative to the current pulse in the amplifier was determined for the most part solely by gas-discharge processes and proved to be poor for small overvoltages. This instability, however, did not affect the results, because in reality it was equivalent to a certain additional shift relative to the current pulse in the amplifier. The amplitude stability of the generator pulse was no worse than 10% and was determined chiefly by the measurement accuracy. The maximum power level of the generator added up to about 50 W over all lines, which is clearly lower than the power at which the amplifier is saturated.

Under conditions such that the population of the lower laser level is small the absolute population of the upper state is easily calculated from the known value of the gain. Inasmuch as the error in the gain measurement is well below the error in the measurements of the absolute population of the upper levels by the comparative method, the information on the population of the laser states could be significantly refined.

§ 3. Gas-Discharge Measurement Technique

We used a low-inductance shunt with a resistance of 0.01 Ω or a Rogowski loop to record the current. The signal from either of these detectors was sent to an oscilloscope through a low-inductance divider, which gave a signal reduction of 1/10 to 1/100. The current was determined from the signal amplitude and shunt resistance measured in the static mode. Under real conditions, due to the skin effect, the current thus measured turned out to be somewhat too high. The absolute value of the current was determined more accurately on the basis of the simple relationship between the charge on the capacitor and the time integral of the current:

$$q = CV = \int_0^\infty J(t)\,dt. \tag{2}$$

The voltage on the working capacitor was measured with an S-96 voltmeter. The potential distribution along the tube and its time variation were recorded by means of electrodes soldered into various locations along the tube, from which the output signal was transmitted by a special switch to a DNE-9 capacitative divider and then to the oscilloscope.

The electron density $n_e > 10^{14}$ cm^{-3} was measured from the broadening of the atomic hydrogen H_β line [59, 60], as well as by an interferometric method [61, 62]. In the first case 5% hydrogen was added to the nitrogen. A special test was made on the basis of the current and spontaneous emission oscillograms to verify that these additives did not introduce any appreciable variation of the pulsed-discharge dynamics. The luminescence of the atomic hydrogen was brought out through a small window in the tube and transmitted to a monochromator, which is used to separate out a part of the broadened atomic line, and then to the oscilloscope. Knowing the half-line width $\Delta\lambda(t)$, we can determine $n_e(t)$:

$$n_e(t) = C(n_e, T_e)\,[\Delta\lambda(t)]^{3/2}, \tag{3}$$

where $C(n, T_e)$ is given in the tables of [59].

This method was used to measure the electron density in N_2 primarily in the afterglow, because at the time of current passage the intensity of the spontaneous emission lines of the 2+ band system of this molecule, which were superimposed on the H_β line, was very large and, of course, distorted the results. The measurement accuracy was limited by the instrument function of the monochromator (0.8 Å) and the comparative imperfection of the mechanical scanning system; the estimated error of the measurements was $\sim 10^{15}$ cm^{-3}.

For the interferometric measurements we used a Rozhdestvenskii interferometer [62, 63]. The light source was a continuous neon–helium laser operating at a wavelength of 0.63 μ. When a plasma with a time-variant refractive index n(t) was inserted into one of the branches of the interferometer, the output of the latter developed intensity beats, which were recorded with a photomultiplier. The mechanical oscillations of the mirrors, at a frequency no greater than $\sim 10^3$ Hz, could be neglected, since the variations of n(t) that interested us took place within a period of ~ 10 μsec. The oscillations of the interferometer mirrors were mainly felt in the initial phase of the intensity beats and did not distort the results.

Both methods of measuring the electron density – the interferometric method and the method based on the broadening of the H_β line – yield mutually consistent results to within $\sim 50\%$.

CHAPTER III

PULSED EMISSION FROM H_2, D_2, AND HD MOLECULES

§1. The Stimulated Emission Spectrum

The explanation of the qualitative inversion mechanism in H_2, D_2, and HD lasers is far simpler than in any other laser of the given type, largely because of the particular simplicity of the stimulated emission spectrum of these three isotopic molecules. Lasing was observed at six lines in hydrogen, at two in deuterium, and at one line associated with HD molecules in a hydrogen–deuterium mixture [5, 10, 36]. All of these lines fall in the spectral interval from 8350 to 13,050 Å. Large-wavelength lasing could not be observed on our equipment, because the detectors had practically zero sensitivity in the system.

TABLE 1. Stimulated Emission Spectrum of H_2, D_2, and HD Molecules at the Electron Transition $^1\Sigma_g^+(E) \rightarrow {}^1\Sigma_u^+(B)$

λ_{air}, A (measured)	ν_{vac}, cm^{-1} (measured)	ν_{vac}, cm^{-1} [54, 55]	Relative intensity[a]	Transition	
				Band	Rotational line
		H_2			
8349.49	11973.49	11973.41	$9\cdot10^{-4}$	(2.1)	P (2)
8876.13	11263.08	11263.11	$1.7\cdot10^{-3}$	(1,0)	P (4)
8898.77	11234.43	11234.37	1	(1.0)	P (2)
11155 [b]	8962 [b]	8956.37	$8.5\cdot10^{-3}$	(0,0)	P (4)
11215 [b]	8915 [b]	8908.51	1	(0.0)	P (2)
13050 [b]	7660 [b]	7656.73 [c]	—	(0,1)	P (4)
		D_2			
8277.53	12077.58	12077.59		(2,0)	P (3)
9529.99	10490.31	10490.30		(1.0)	P (3)
		HD			
9160 [b]	10914			(1.0)	—

[a] Conditions: hydrogen pressure, 2.1 mm Hg; voltage 26 kV; repetition rate, 8 Hz; transmissivity of silvered mirrors, 0.25 and 12%; intensity of the P(2) (1, 0) line adopted as unity in the (1, 0) and (2, 1) bands; intensity of the P(2) line adopted as unity in the (0, 0) band.

[b] Wavelengths measured with ~10 Å error on a monochromator.

[c] Calculated from energy level tables in [55].

The results of our investigation of the stimulated emission spectrum are summarized in Table 1. The measured wave numbers of the stimulated emission lines are compared with the frequencies of the spontaneous emission lines, which are known from [54, 55] to within a few hundred reciprocal centimeters (third column). As the table indicates, the discrepancy between our measurements and the data of [54, 55] does not exceed 0.08 cm^{-1}. This fact enabled us to allocate all the stimulated emission lines with precisely measured wavelengths to their appropriate molecular transitions, which are given in the last column of the table. The left half of this column gives the band, and the right half the rotational line. For the laser emission lines marked with asterisks the postulated transition allocation is given in the table. Also presented in the table are the relative intensities of the emission lines. The fact that lasing from the isotopic H_2 and D_2 molecules is observed at the E–B transition affords stronger support for the validity of the given transition allocation.

Unlike spontaneous emission, which can be observed at a great many vibrational transitions [54, 64], stimulated emission, i.e., lasing, is observed only at transitions from low ($v' \le 2$) vibrational levels of the E state to low ($v'' \le 1$) vibrational levels of the B state. On the other hand, the strongest lasing bands for hydrogen begin at the v = 0, 1 level [(1, 0) and (0, 0) bands]; for deuterium the (2, 0) and (1, 0) bands are predominantly excited with initial levels v' = 2, 1, while the (0, 0) band starting with the v' = 0 level is generally not observed. The first thing that catches the eye with respect to the rotational structure is the fact that lasing is observed only from the ortho modifications of the hydrogen and deuterium molecules. The strongest stimulated emission lines for hydrogen begin at the J' = 1 level, while emission from the J' = 3 level is weaker by three orders of magnitude. For deuterium lasing is observed only from the J' = 2 level. The stimulated emission spectrum discloses only the P branch, whereas in spontaneous emission at the transition $^1\Sigma_g^+ \rightarrow {}^1\Sigma_u^+$, according to the selection rules [65], the R branch can also be observed.

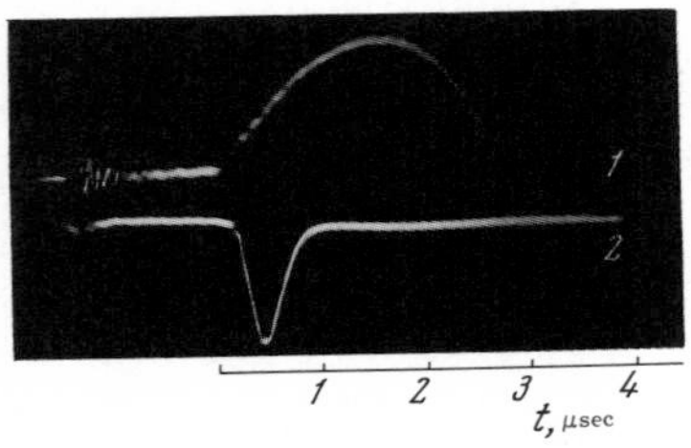

Fig. 3. Oscillogram of current (1) and stimulated emission (2) pulses in H_2. Voltage, 13.4 kV; p = 0.7 mm Hg; discharge tube bore, 1.5 cm; tube length, 2.0 m; mirror transmissivities, 1 and 17%.

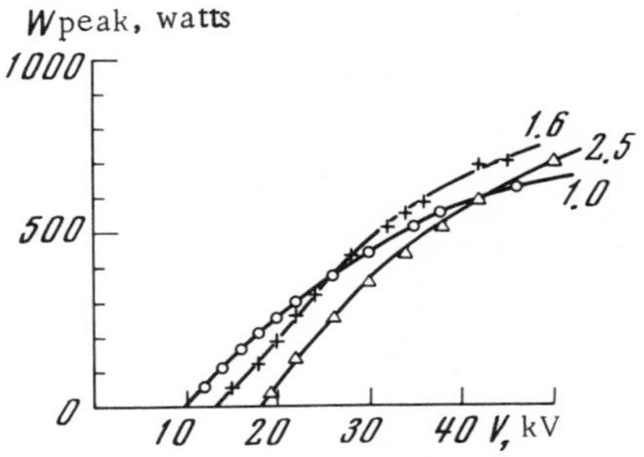

Fig. 4. H_2 peak laser power versus voltage at various gas pressures. Tube bore, 3.0 cm; length, 1.3 m; f = 10 Hz; C = 0.01 μF; mirror transmissivities, 1 and 17%. The hydrogen pressure in mm Hg is given alongside each curve.

The laser action of molecular hydrogen, like the far weaker laser action of D_2 and HD molecules, is observed as a short pulse of width ~ 0.2 μsec at the leading edge of the current pulse. The latter fact is especially typical of the laser action of these three molecules. Typical oscillograms of the stimulated emission and current pulses are shown in Fig. 3. The optimum pressure for lasing depends slightly on the voltage on the tube, amounting to about 3 mm Hg at 26 kV (tube length, 1.3 m). With an increase in voltage the peak laser power at first climbs rapidly, then begins to level off (Fig. 4). A change in the excitation-pulse repetition rate does not produce any variation of the laser power.

The maximum attainable laser power is about 0.7 to 1.0 kW. At the optimum mirror transmissivity of ~ 20% a large part of the laser power is concentrated at the P(2) line of the (1, 0) band (λ = 8898.77 Å). Approximately 20% of all the lasing energy belongs to the P(2) line of the (0, 0) band (λ = 11,222.05 Å). The remaining lines are two or three orders of magnitude weaker. We note that with the proper choice of mirrors it is easy to obtain lasing with practically all its power concentrated either at the P(2) (1, 0) line or at the P(2) (0, 0) line. The gain at the strongest line, P(2) (1, 0) attains 0.01 cm^{-1}.

Sample oscillograms of the current, molecular spontaneous emission near λ = 8898.77 Å, and spontaneous emission at the H_α line of atomic hydrogen are given in Fig. 5 (curves 1-3, respectively). The stimulated emission pulse has a considerably shorter duration (at the half-peaks) than the spontaneous emission pulse, but their maxima are attained at about the same time. Amplification experiments show that inversion exists virtually the entire time that spontaneous emission is observed (curve 6).

In low-loss cavities under certain excitation conditions a stimulated emission pulse of complex shape and width to 5 μsec is observed (Fig. 6). Notice the slight "pedestal" at the beginning of both pulses. The height of the pedestal can attain 0.1 times the height of the emission peak. As a rule, a considerable part of the lasing energy belongs to the short peak that appears with the transmission of strong currents through the tube. In cavities with high-transmission mirrors the pedestal is not observed.

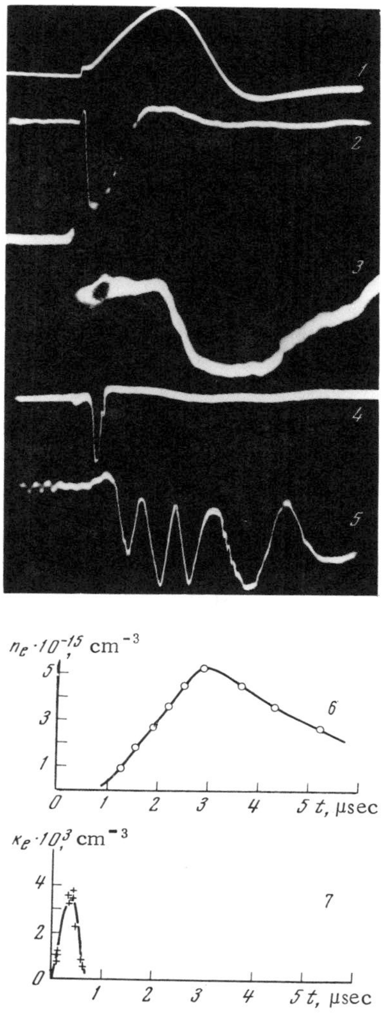

Fig. 5. Dynamics of spontaneous emission in H_2. 1) Current; 2) spontaneous emission at molecular transitions; 3) the same, at the H_α line; 4) stimulated emission pulse; 5) oscillogram of beats at the output of a Rozhdestvenskii interferometer; 6) electron density calculated from oscillogram 5; 7) gain. Gas pressure, 0.75 mm Hg; voltage, 11.7 kV; C = 0.11 μF.

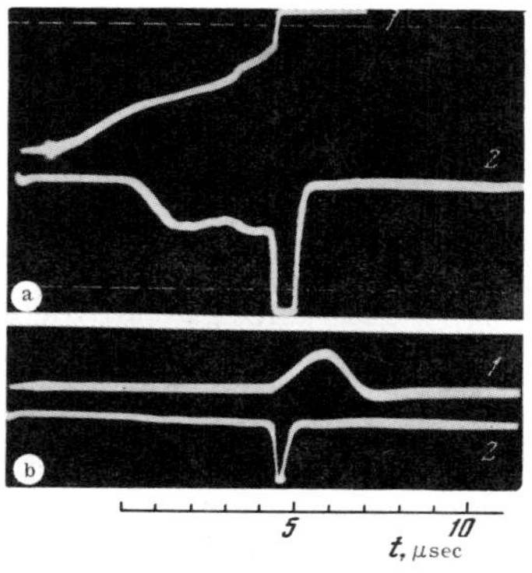

Fig. 6. Oscillograms of a "long" stimulated emission pulse in H_2. 1) Current; 2) stimulated emission. In going from (a) to (b) only the sensitivity of the current- and stimulated-emission-recording channels is changed. The horizontal trace indicates the linear-deflection limit of the oscilloscope beam. Transmissivity of dielectric mirrors, 1%; p = 0.7 mm Hg; C = 0.11 μF.

§ 2. Electron States and Transitions in H_2, D_2, and HD Molecules; Qualitative Inversion Mechanism

The general pattern of the potential-energy curves for the upper and lower working electron states and electron ground state of a molecular hydrogen laser is illustrated in Fig. 7. The vibrational levels are marked by horizontal lines. The bands observed in stimulated emission are indicated by short vertical arrows. The ground-state potential curves of molecular hydrogen, designated summarily by the letter (X), and the lower laser state (B) have been calculated in [66]. Note the strong shift of the potential curves for these states relative to one another. The equilibrium internuclear separations differ by a factor of about 1.7.

The potential curve for the upper laser state has been calculated theoretically in [67]; it has two minima. In the vicinity of the inner minimum the wave function of this state is close to the wave function of the $2s\sigma^1\Sigma_g^+$ state, and in the vicinity of the outer minimum it is close to the wave function of the $(2p\sigma)^2\,{}^1\Sigma_g^+$ state (indicated by arrows in Fig. 7). The inner minimum corresponds to the experimentally observed E state [68] and has direct bearing on the problems of interest here. As for the minimum corresponding to the experimentally observed $(2p\sigma)^2\,{}^1\Sigma_g^+$ state, its role is largely secondary as far as we are concerned. The potential curve for the E state is shifted relative to the ground state, but is considerably less so than the B state.

Of the other electron states, it is important to mention the $2p\pi^1\Pi(C)$ state, whose potential curve practically coincides in the lower part with the E-state curve. The fundamental parameters of these two states are very similar [65]. The potential-energy curves for the isotopic molecules D_2 and HD, as we are aware [65], are practically the same as for molecular hydrogen.

The radiative transitions and selection rules are the same for H_2, D_2, and HD molecules. Transitions from the upper laser state to the ground state are forbidden. The lower laser state and the C state are optically coupled to the ground state. In this case the Lyman and Werner bands, which lie in the vacuum ultraviolet region, are respectively observed. In the infrared region the spectrum of the E–B electron transition and its associated lasing are observed. The C–B transition is forbidden, and the E–C transition is allowed. The latter transition is in the vicinity of 80 to 100 μ and was not observed experimentally (see [65] for more details on the transitions and selection rules).

It is a simple matter to calculate [70] the radiative lifetimes of the B and C states from the known [70, 71] oscillator strength f_e for X–B (f_e = 0.2) and X–C (f_e = 0.4) transitions:

$$\tau_r = \frac{1}{f_e}\cdot 1.51\cdot\frac{G'}{G''}\frac{\nu_{abs}}{(\nu_{em})^3}, \tag{4}$$

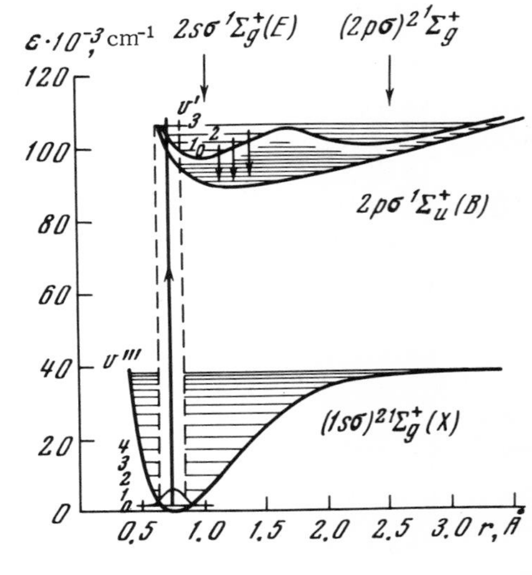

Fig. 7. Potential-energy curves for molecular hydrogen. Abscissa: internuclear separation; ordinate: energy. In the lower part of the potential curve for the X state is given the square of the wave function for the v''' = 0 vibrational state. We use the symbol v''', after [69], to denote the ground-state vibrational levels.

where G' and G'' are the statistical weights of the upper and lower electron states, ν_{em} and ν_{abs} are the emission and absorption transition wave numbers determined from the position of the potential curves [70].

The radiative lifetimes of the B and C states are equal to 1.6 and 0.5 nsec, respectively. The literature does not contain any information on the probability of spontaneous emission at the working B–E transition or the oscillator strength of this transition.

In view of the fact that pulsed laser action occurs in H_2, D_2, and HD at the leading edge of the current pulse, it seemed natural, in order to explain this action, to invoke the "direct electron impact" mechanism that we proposed in [2] for interpreting the emission spike in neon. This mechanism is based, on the one hand, on the hypothesis of direct excitation of the working levels by ground-state electrons and, on the other, on a consideration of the fact that such excitation is realized under nonsteady conditions. The general idea of the mechanism is as follows. In the first instants of buildup of pulsed discharge the molecular states are filled primarily by virtue of inelastic collisions of electrons with molecules concentrated in the ground state. Inasmuch as the effective cross sections for excitation of different molecular states by electrons are almost always different, the filling process is nonuniform. This fact accounts for the onset of inversion at the leading edge of the current pulse. If excitation is fast enough, radiative decay can be neglected. In this case inversion takes place in a three-level system under the condition that the excitation rate of the upper laser level from the ground state is higher than that of the lower level. The ensuing inversion lasts for a certain period of time determined by the relaxation time of the upper laser level and by the development of step-by-step filling processes, which subsequently lead to the cessation of inversion as the discharge develops.

Under conditions such that the dependence of the effective cross sections for excitation of the laser states on the electron velocity is roughly identical (as is true in the H_2 laser) it is necessary, in order to obtain a higher filling rate for the upper laser state, that the effective excitation cross section for this state be larger in absolute value than that of the lower state. The data on the excitation cross sections for the B and E states are as follows: In the vicinity of energies ~ 100 eV the effective excitation cross section for the upper laser state is approximately one third the cross section for the lower laser state [72]. The latter ratio of the cross sections would seem, in general, to deny population inversion. It must be realized, however, that the data of [72] refer to all vibrational levels of the electron state of the molecule as a whole. The excitation of individual vibrational levels within the limits of a given electron state, on the other hand, can exhibit extreme variability. Undoubtedly, it is required to take account of the specific attributes of the vibrational-level excitation.

The inception of inversion in the H_2 laser has the following qualitative pattern (only the details differ for D_2 and HD lasers).

In connection with the large interpulse times and relatively low repetition rate of the experimental pulsed discharge (current pulse width, ~ 1 μsec; repetition rate, up to 20 Hz), at the beginning of the next current pulse practically all the molecules are concentrated in the ground state. Due to the large vibrational and rotational constants at room temperatures (~ 350°K) only a few of the lower rotational levels of the v''' = 0 vibrational state are populated, and it is from these levels that the "empty" upper levels begin to fill up:

$$H_2(X_0) + e \nearrow H_2(E_{v'}) + e, \quad (5)$$

$$H_2(X_0) + e \searrow H_2(B_{v''}) + e, \quad (6)$$

In processes (5) and (6) only a few vibrational levels of the laser states are rapidly populated. This fact is attributable to two causes. On the one hand, the excitation rate of the $E_{v'}$ level (the same applying to the $B_{v'}$ level) is proportional to (see, e.g., [73]) the Franck–Condon factors, which are generally largest for transitions preserving the internuclear separation. Since the ground-state molecules are concentrated in a relatively small interval of internuclear separations (v''' = 0), it follows from (5)-(6) that not all, but only some vibrational levels of the B and E states are excited. On the other hand, since the equalizing action of vibrational relaxation is certainly small [74, 75], the nonuniformity of the vibrational-level populations is maintained. From the standpoint of lasing the favorable cases are those in which the upper laser level, as the one most abundantly populated in process (5), has a fairly high probability of transition to the lower level, which is weakly excited in (6). Such cases can be detected from the position of the potential curves, on which the Franck–Condon factors and, hence, the excitation rates of the laser levels are strongly dependent. The analysis of the inversion mechanism between vibrational levels by means of the potential curves and by the Franck–Condon principle is distinguished by great physical transparency and at the same time leads to qualitatively valid results.

According to the Franck–Condon principles, the molecules concentrated at the v''' = 0 level of the ground state are excited by electron impact "vertically," i.e., without any change in the internuclear separation or velocities of the nuclei. In this case only those vibrational levels of the B and E states that fall within the compass of the vertical band indicated in Fig. 7 are predominantly populated. The potential curves for the B and E states are situated such that the lower vibrational $E_{v'}$ levels (v' = 0, 1, 2) are effectively populated in process (5). Due to the strong shift of the potential curve for the B state relative to the ground state the vertical band falls at levels v'' $\geq$ 3. But the excitation of the lower vibrational levels v'' = 0, 1, 2 of this state by direct electron impact is unlikely in process (6), because the Franck–Condon principle is contradicted. The probabilities of transition from the $E_{v'}$ level (v' = 0, 1, 2) to the $B_{v''}$ level (v'' = 0, 1, 2), judging from the shape of the potential curves, must be relatively high. In accordance with the foregoing discussion inversion should be expected at transitions from the lower levels of the $E_{v'}$ state (v' = 0, 1, 2) to lower levels of the $B_{v''}$ state (v'' = 0, 1, 2). The corresponding transitions are indicated by arrows in Fig. 7. We have thus explained one of the observed characteristics of the vibrational structure of the emission spectrum, namely the fact that laser emission is observed only at a few of the bands observed in spontaneous emission. Moreover, we can predict exactly the transitions at which stimulated emission must occur, and indeed it was observed experimentally at transitions from the lower vibrational levels of the E state to lower vibrational levels of the B state.

On the basis of the postulated inversion mechanism we also gain insight into the differences in the vibrational structure of the laser emission spectrum of H_2 and D_2 molecules, in particular the fact that lasing is not observed in deuterium from the vibrational v' = 0 level, whereas very strong lasing is observed from this level in hydrogen. The following consideration needs to be taken into account.

The position of the vertical band in which molecular excitation occurs is the same for all three molecules, because the shape of their potential curves is the same. However, due to the isotopic shift the energy levels of the D_2 and HD molecules are moved downward relative to the hydrogen molecule, the deuterium molecule exhibiting the greatest shift. As the $E_{v'}$ levels of the isotopic molecules "slip" downward their excitation conditions change. Whereas for hydrogen, for example, the E_0 level falls inside the excitation band and exhibits very strong lasing action, for deuterium this level drops so far down as to almost fall out of the excitation band, and any lasing from it must be either extremely weak or not be observed at all. Consequently, laser emission from the E_0 level of deuterium was not observed in the experiments.

We see, therefore, that even our relatively simple considerations concerning the inversion mechanism on the basis of the Franck–Condon principle and the potential-energy curves give a qualitatively realistic picture of the excitation of the vibrational structure of the laser emission spectrum for hydrogen, as well as the modifications of that structure in transition to the isotopic molecule D_2.

§3. Singular Characteristics of the Excitation of Rotational Spectral Structure in Hydrogen and Deuterium Lasers

We now consider how the "direct electron impact" mechanism can be used to account for the observed characteristics of the rotational structure of the laser emission spectrum. By contrast with other presently known lasers acting on diatomic molecules, only H_2, D_2, and HD lasers offer the possibility of analyzing the details of the excitation of individual rotational levels.

As an example we consider the filling of the rotational levels of the (1, 0) band of molecular hydrogen and then elaborate on the attributes of the excitation of the deuterium levels [76]. The laser-level diagram for the (1, 0) band is given in Fig. 8.

In comparison with experiment it is useful to have some information on the population distribution with respect to the rotational levels of the upper and lower states. In this case we can gain at least a qualitative notion of the stimulated emission intensity distribution among the rotational lines [77].

The formation of inversion in the initial stage of pulsed discharge development, as viewed within the scope of the "direct electron impact" mechanism, automatically assumes that the molecular gas has not yet had a chance to become significantly heated by elastic collisions with electrons. This assumption connotes that the distribution of the molecules among the rotational levels of the ground state must correspond to the average temperature established in the discharge tube in the current-pulse repetition mode. Under actual conditions the average temperature of the tube did not exceed 320°K. At this temperature the J = 1 level is the most vigorously filled (65% of all the molecules). The population of the J = 3 level (8%) and J = 0 and J = 2 parahydrogen levels (each about 12%) is considerably lower, while the population of J ≥ 3 levels can

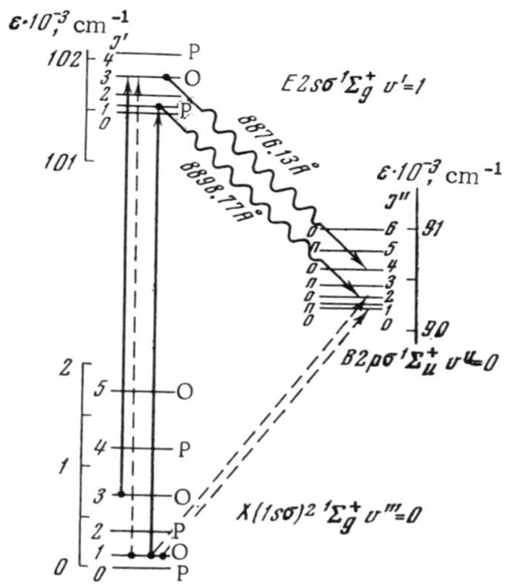

Fig. 8. Laser-level diagram of the electronic vibrational states X_0, B_0, and E_1 of the hydrogen molecule. The rotational levels are drawn to scale. The distance between electron levels is arbitrary. J) Rotational quantum number; O) orthohydrogen; P) parahydrogen; v) vibrational quantum number. The diagram was compiled from the H_2 energy level tables of [55].

be neglected. Of all the molecules in the vibrational ground state, 75% belong to orthohydrogen in accordance with the statistical-weight ratio of the nuclei: 3:1.

We recall that in electronic excitation the effective excitation cross section of a molecule with preservation of J is considerably greater than for excitation with variation of J, and transitions between modifications are virtually excluded. This means that excitation from each vibrational level of the ground state goes primarily to one rotational level of the upper state (if the condition $\Delta J = 0$ can be met), and the filling up of the ortho and para modifications occurs independently. Thus, the population rate of the individual rotational levels of the E state must to a certain extent correspond to the population of the ground-state rotational levels. The most rapidly populated is the $E_1 1$ level, the excitation of the $E_1 3$ level being necessarily much weaker. Inasmuch as only a fourth of all the molecules are attributed to the para modification, it is clear that the population rates of the $E_1 0$ and $E_1 2$ levels must also be very small.

Population "mixing" of the rotational levels of the E state due to rotational relaxation must, all things considered, be very small. The establishment of an equilibrium distribution among the rotational levels of the ground state requires about 300 collisions [74, 75]. Under our conditions (hydrogen pressure, 2.5 mm Hg; temperature, 320°K) this implies a relaxation time of 10^{-5} sec, which is about an order of magnitude greater than the inversion holding time in a high-current discharge.

Under these conditions the population distribution among the rotational levels of the upper working state and ground state must be largely similar. As for the population of the lower laser state, in the amplification regime at any rate, it must be negligibly small. This result is attributable, on the one hand, to the fact that the rate of population of this state with ground-state electrons is far lower than for the upper laser levels, because it involves violation of the Franck–Condon principle. On the other hand, account must be taken of the exceedingly small radiative lifetime of the lower laser levels, which decay in the vacuum ultraviolet region, whereas transitions from the E state fall in the infrared.

Taking all of the foregoing into account, along with the fact that the emission probabilities in transition from one rotational level to another change rather slowly, we clearly perceive that not only population inversion and amplification, but also the laser emission intensity must qualitatively correspond to the population distribution among the ground-state rotational levels. For example, the P(2) line of the (1, 0) band, corresponding to the most populated $X_1 1$ level, must have a far greater power than the P(4) line, which corresponds to the much less populated $X_1 3$ level. In view of the fact that the rotational structure of different vibrational levels of the B and E states is practically the same, in any band laser emission must be observed first and foremost from the $J' = 1$ level of the ortho modification.

These qualitative conclusions based on the "direct electron impact" mechanism of excitation of the laser levels are fully supported by the experimental results. Thus, in the (1, 0) band the P(2) line yields the strongest laser action, and the P(4) line the weakest action. At all other bands of the H_2 laser, emission is observed for the most part specifically at the P(2) lines.

Everything said so far with regard to molecular hydrogen can be translated almost verbatim to its isotopic partner D_2. Under isotopic substitution there are well-defined variations in the relative population of the ground-state rotational levels. At a temperature of ~ 320°K the $J = 2$ level of orthodeuterium is the most abundantly populated (40% of all molecules). The remaining levels are less populated. Judging from this population distribution in light of the foregoing discussion, one should expect lasing chiefly from the $J' = 2$ level for hydrogen. In fact (see Table 1), in the (1, 0) band lasing was observed only at the P(3) line with starting level $J' = 2$. Similarly, in the (2, 0) band the only laser emission line observed was P(3). Consequently, the variations in the rotational structure of the laser emission spectrum under isotopic substitution can be qualitatively predicted. The fact that lasing was observed only in the P branches can be explained by realizing that the probability of radiative transfer to the P branches of the electron transition ${}^1\Sigma_g^+(E) \rightarrow {}^1\Sigma_u^+(B)$ is greater than to the R branches [the P(2) lines have twice as high a transition probability as the R(2) lines].

About six months after our paper [5] was published, lasing in H_2, D_2, and HD was also observed in [37]. Six emission lines were additionally observed with wavelengths $\lambda > 1.2\,\mu$, above the long-wave sensitivity of our recording-detection system. The inception of lasing at these lines is fully explained on the basis of the "direct electron impact" mechanism.

We now consider a few uncertain problems. It is not entirely clear, for instance, what is the role of the C state in the population and de-excitation of the upper laser state, with which it is in perfect resonance. The electron energies of the E and C states attain an absolute value of 10^5 cm^{-1}, differing only 20 cm^{-1} from one another. Besides the coincidence of the energy levels, the potential curves are also very nearly congruent. The equilibrium distances differ only by 0.02 Å. It is reasonable to suppose that under these conditions, by virtue of collision between molecules and between electrons and molecules, a transfer of excitation energy between the E and C states is possible. The direction of this process depends on the population of the states. The relative population of the C state must be small on account of the exceedingly rapid radiative decay of this state ($\tau_r = 5 \cdot 10^{-10}$ sec). Under these conditions the excitation energy will transfer predominantly from the E to the C state with the subsequent emission of an energy quantum in the vacuum ultraviolet. This process must lead to a certain reduction in the population of the E state.

Another problem not entirely clear is the lasing cutoff mechanism. Some data indicate that cutoff is unrelated to the accumulation of molecules in the lower laser state. This conclusion seems wholly sensible if we take account of the extremely short radiative lifetime of this state. Moreover, appreciable absorption at the working transition was not observed in the experiments, and lasing was observed practically the whole time that there was appreciable spontaneous emission at molecular transitions.

All of this, combined with the nature of the molecular de-excitation (see Fig. 5), leads to the conclusion that lasing cutoff occurs due to a reduction in the population of the upper laser state during the discharge buildup process. This reduction can be attributed to a whole series of causes. Evidently, the main factor is dissociation. We are aware [28] that hydrogen molecules can be excited by electrons in an unstable state with subsequent dissociation into two atoms. The cross section for this process is very large, $4 \cdot 10^{-17}$ cm^2. Straightforward estimates indicate that for $n_e \sim 10^{15}$ cm^{-3} the time required for dissociation of a third of all the molecules is about 0.25 μsec. Thermal dissociation due to rapid heating of molecular hydrogen is also entirely possible. Especially strong heating is expected for exactly this molecule, as the lightest of the diatomic molecules. In elastic collisions with electrons the hydrogen molecule acquires a fractional energy proportional to the electron–molecule mass ratio, which is a maximum for H_2. In elastic collisions with protons and H_2^+ ions all the energy spent by the field in the acceleration of protons is imparted to the hydrogen molecules. Due to the large drift velocity involved, this energy is also very much larger than for other diatomic molecules. Finally, hydrogen is heated further in the excitation of unstable molecular states, which decay into two atoms with a sizable release of kinetic energy (4 to 8 eV). This entire energy is used in heating the molecules.

As a result of the stated processes, for $n_e \sim 10^{15}$ cm^{-3} hydrogen can become rapidly heated to several thousand degrees, resulting in vigorous thermal dissociation of the hydrogen molecules. Together with the dissociation induced by collisions with electrons, the foregoing process will necessarily evoke a sharp reduction in the total number of molecules and, as a result, a reduction in the spontaneous emission of molecules, as is indeed observed experimentally.

The following consideration must also be brought to attention. As a result of dissociation the number of particles with which electrons can enter into inelastic collision is sharply increased. Under conditions such that the field remains constant or even increases (due to discharge of the capacitor), the indicated process cannot help but diminish the average electron energy. As a result, there must occur a sharp reduction in the excitation rate of the molecular and atomic states. If this fact is ignored, it is difficult to account for the experimentally observed time variations of the intensity of the H_α line of atomic hydrogen (see Fig. 5). As evident from Fig. 5, despite the abrupt increase in the electron density the intensity of the H_α line not only does not increase, it even decreases slightly, until such time as the contribution of recombinational filling becomes significant.

One more problem to be scrutinized is the stimulation of relatively long-duration (~ 5 μsec) laser emission in H_2 (see Fig. 6). This emission, whose power is estimated to attain one or more watts, is observed under conditions such that the current is essentially unchanged. Due to the relative invariance of the excitation conditions it seems unlikely that the stimulation of emission would be related to nonstationary processes, whose duration is normally much shorter than 5 μsec. The observed situation is clear evidence of the hypothesis advanced in [37] regarding the possibility of continuous lasing in H_2. This possibility is largely supported by the extremely short radiative lifetime of the lower laser state.

The set of problems spelled out above, which remain shrouded in a certain obscurity, pertain more to the subtler details of inversion than to its basic substance. The notions of direct excitation of the laser levels by "direct electron impact," augmented by the Franck–Condon principle, afford a fairly comprehensive qualitative picture of the formation of inversion in H_2, D_2, and HD lasers. This conception was subsequently used to give a qualitative interpretation of pulsed lasing for all presently known pulsed gas lasers operating on electron transitions in diatomic molecules, including the $N_2(1+)$ laser [18, 23]. An analysis of the operation of the latter is given in Chapter IV.

CHAPTER IV

TEMPORAL AND ENERGY CHARACTERISTICS OF THE FIRST-POSITIVE-BAND-SYSTEM MOLECULAR NITROGEN LASER

As in the case of hydrogen, lasing in the 1+ band system of N_2 is observed at the leading edge of the current pulse. Sample oscillograms of the stimulated emission and current pulses for low and high voltages, of the type shown in Fig. 9, bear out this conclusion. The low-voltage emission pulse has the usual characteristic shape, i.e., a relatively gradual leading edge and sharp cutoff. It is clear that when the voltage is increased the laser emission pulse width Δt_L is sharply curtailed, by about an order of magnitude under the conditions of Fig. 9. Also decreased is the time interval t_{max} in which the maximum laser power is attained. The voltage dependence of Δt_L and t_{max} is shown in Fig. 10. Both of these variables decrease rapidly as the voltage is increased. Under these conditions the laser emission pulse energy varies somewhat differently with increasing voltage than the peak power (Fig. 11); after reaching a maximum, the energy curve drops off more rapidly than the laser power.

The dependences illustrated in Figs. 10 and 11 are the most pronounced for relatively low cavity losses (~ 1 to 3%), as well as for long tubes (~ 2 to 6 m) and reasonably large capacitances (≲0.03 μF). However, a low transmissivity of the mirrors automatically lowers the laser output power. For optimum mirror transmissivity (~ 20%), relatively short tubes (1.3 m), and a capacitance of 0.01 μF, as in most of the measurements of the laser characteristics, the lasing

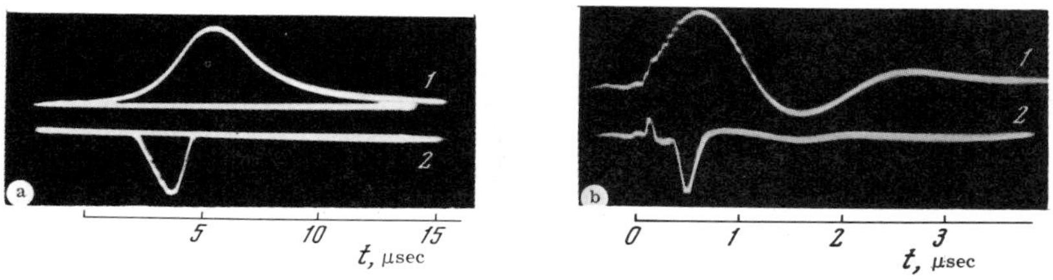

Fig. 9. Oscillograms of current (1) and stimulated emission (2) pulses. Voltage: a) 9.6 kV; b) 30 k; p = 0.75 mm Hg; tube length, 2.0 m; bore, 1.5 cm; C = 0.03 μF; transmissivity of dielectric mirrors, 1%.

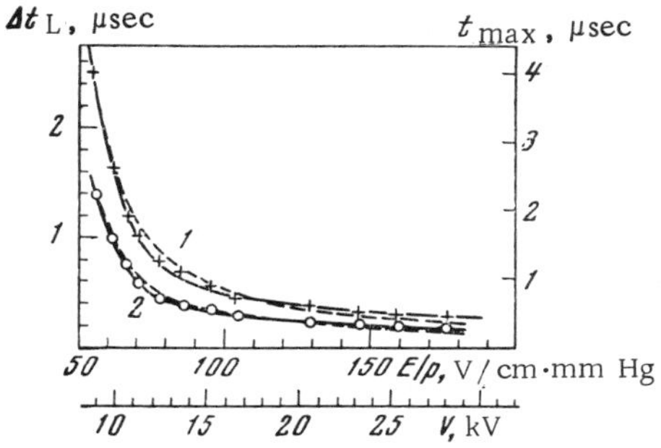

Fig. 10. Time t_{max} at which peak laser power is attained (curve 1) and laser emission pulse width Δt_L (curve 2) versus voltage under low-loss cavity conditions. Calculated dependence represented by dashed curves (see Chapter VIII); experimental conditions the same as in Fig. 9; extra scale for the parameter E/p = ratio of field to pressure, to facilitate comparison with analytical curve; time t_{max} measured from beginning of high-current phase of pulsed discharge (see Chapter VI).

Fig. 11. Peak power inside (W^*_{pk}) and outside (W_{pk}) the cavity (curve 1), and laser emission pulse energy (curve 2) versus voltage for a low-loss cavity. Experimental conditions the same as in Fig. 9.

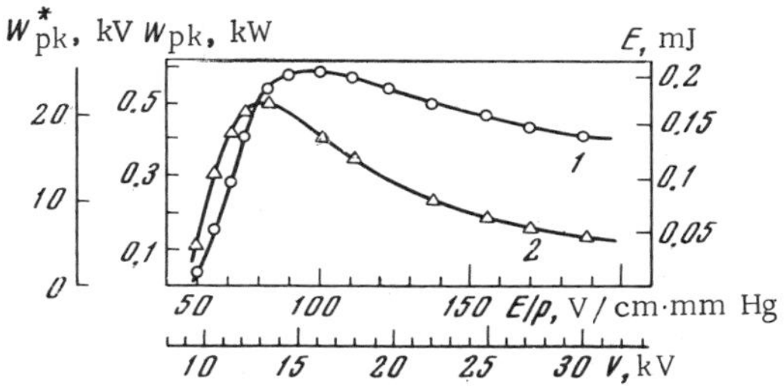

period remains roughly constant at the ~ 0.2-μsec level, exceeding it only near the threshold. The laser energy and peak power have approximately the same type of dependence on the voltage under these conditions. Typical characteristics are shown in Fig. 12. With an increase in pressure the maximum laser power is attained at increasingly higher voltages. The interval of attainable pressures (up to about 8 mm Hg) was restricted by the limiting voltage that could be handled by the high-voltage rectifier (~ 50 kV). Over this entire pressure range the attainable laser power increases approximately linearly with the pressure (Fig. 13).

The laser power increases with the tube bore diameter as well. This is already evident from a comparison of Figs. 12a (bore = 2.0 cm) and 12b (3.0 cm). In a tube of bore 3.0 cm and active length 1.3 m (see Fig. 12b) a power of 20 kW is attained, which is more than two orders of magnitude greater than any power previously announced in the literature for an $N_2(1+)$ laser. Investigations carried out for tubes of various bore diameters, from 0.7 to 4.5 cm, show that at a fixed pressure the peak power increases roughly as the cross section of the discharge tube.

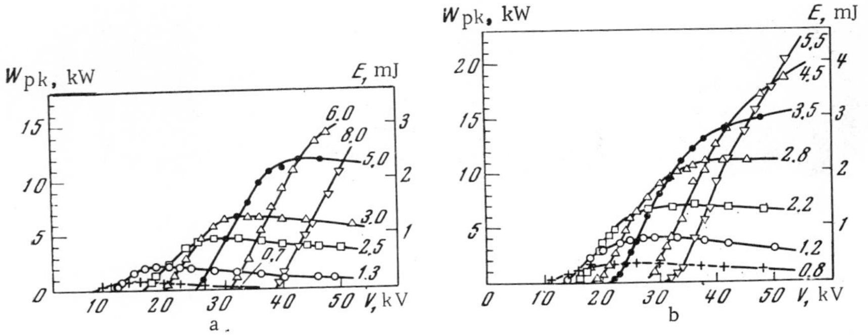

Fig. 12. Peak power W_{pk} and laser pulse energy E versus voltage at various pressures. The numbers attached to the curves indicate the pressure (in mm Hg). Total mirror transmissivity, 20%; a) tube bore, 2.0 cm; b) 3.0 cm; active length, 1.3 m; C = 0.01 μF; f = 10 Hz; peak power computed from measured energy of laser pulse and pulse width (0.2 μsec).

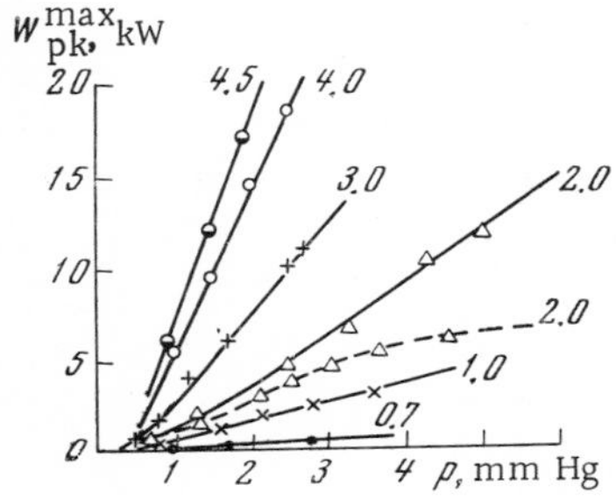

Fig. 13. Maximum value of peak laser power versus pressure for various tube bores. The numbers attached to the curves indicate the tube bore (in cm).

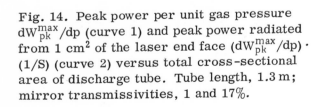

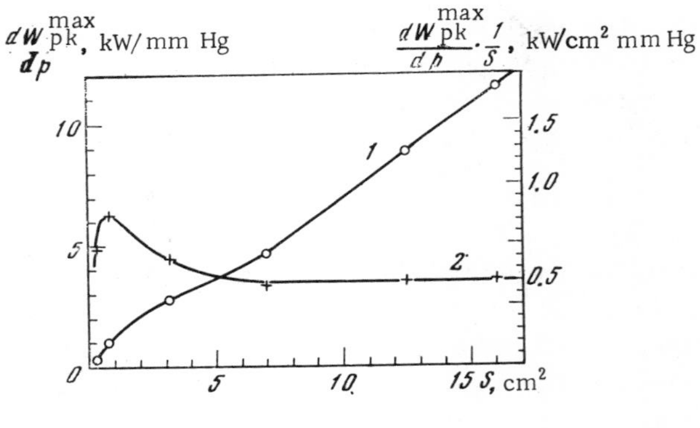

Fig. 14. Peak power per unit gas pressure dW_{pk}^{max}/dp (curve 1) and peak power radiated from 1 cm^2 of the laser end face $(dW_{pk}^{max}/dp)\cdot(1/S)$ (curve 2) versus total cross-sectional area of discharge tube. Tube length, 1.3 m; mirror transmissivities, 1 and 17%.

The slope of the curves $W_{pk}^{max} = f(p)$ increases approximately linearly with the tube cross section (Fig. 14, curve 1), while the quantity $(dw_{pk}^{max}/dp)(1/S)$ essentially stays constant for $S > 5$ cm^2 (curve 2). The peak power distribution over the discharge tube cross section is more or less uniform, particularly at elevated repetition rates (Fig. 15). We note that increasing the tube diameter above the optimum in continuous gas lasers, as a rule, only lowers the peak laser power [78, 79].

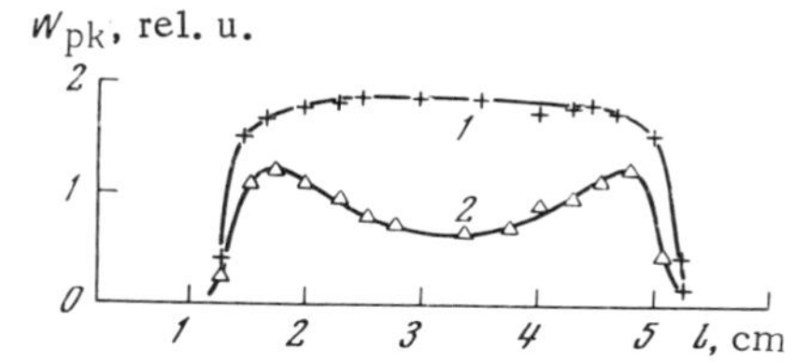

Fig. 15. Lasing intensity distribution over end face of wide tube at repetition frequencies of 20 Hz (curve 1) and 2 Hz (curve 2). Tube bore, 4 cm; p = 1.3 mm Hg; C = 0.01 μF.

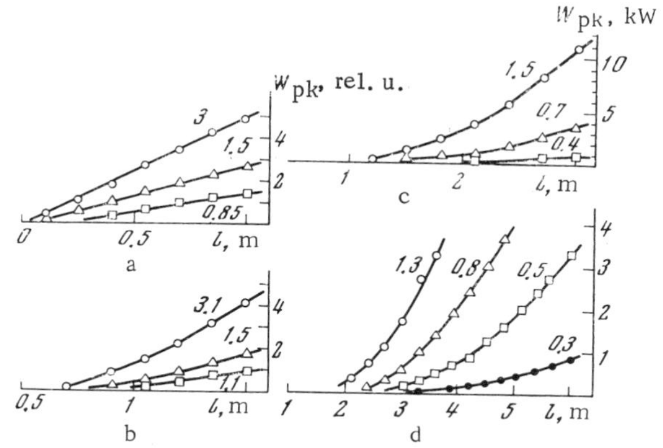

Fig. 16. Peak power versus discharge gap length for various cavity losses and tube bores. The numbers alongside the curves indicate the nitrogen pressure (in mm Hg). a and b) tube bore, 7 mm; a) losses, 2%; b) 20%; c and d) tube bore, 2.5 cm; c) without intermediate mirrors; d) with aluminized intermediate mirrors.

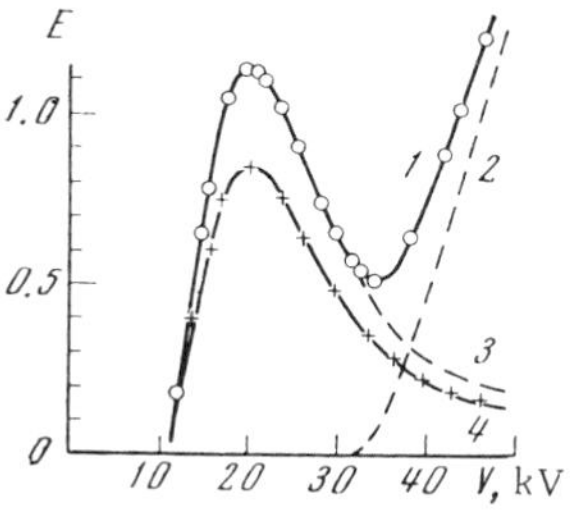

Fig. 17. Relative laser pulse energy in the infrared (curve 3) and ultraviolet (curve 2) band systems of molecular nitrogen versus voltage. 1) Total energy of laser pulse; 4) infrared laser energy separated out by a filter. Tube bore, 7 mm; length, 1.5 m; gas pressure, 2.1 mm Hg; transmissivities of aluminized mirrors, 0 and 4%.

Besides the tube bore and pressure, increasing the length of the discharge gap can contribute significantly to the enhancement of the peak power (Fig. 16). With an increase in the cavity losses (from ~ 2% in Fig. 16a to ~ 80% in Fig. 16d) the power is observed to increase at larger and larger active lengths (starting with 0.6 m and about 2.0 m, respectively). As the pressure is increased, on the one hand, this initial length decreases, while, on the other hand, the slope of the power–length curve increases. As a result, the peak power rapidly increases with the gas pressure and length of the discharge tube.

In connection with the fact that the upper working levels of the N_2(1+) laser are simultaneously the lower working levels for high-power lasers operating at the ultraviolet (2+ band system) transitions of the same molecule, it seems logical to inquire to what extent stimulated ultraviolet emission affects the characteristics of infrared emission. Concurrent lasing in both of these systems is readily observed in narrow tubes of bore 0.5 to 1.0 cm. As apparent from Fig. 17, lasing at the infrared transitions (curve 3) begins at considerably lower voltages (about one third) than at the ultraviolet transitions (curve 2). Despite the rapid growth of lasing energy in the ultraviolet at high voltages, the lasing intensity in the infrared falls off abrupt-

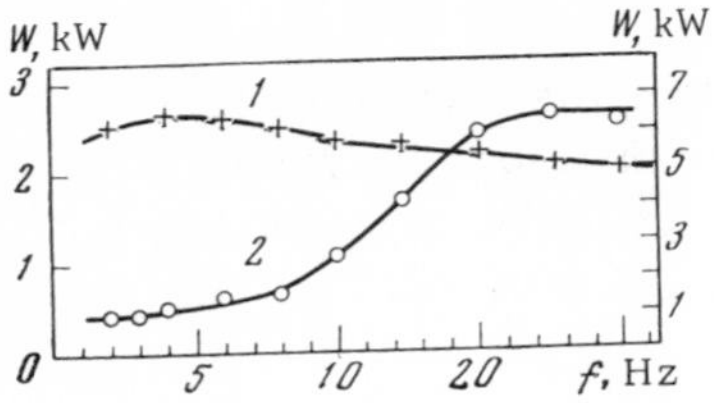

Fig. 18. Peak power versus repetition rate. 1) Gas pressure: 1.8 mm Hg, V = 26 kV (left ordinate axis); 2) gas pressure: 5.5 mm Hg, V = 42 kV (right ordinate axis). Discharge tube bore, 2.0 cm; length, 1.3 m; C = 0.01 μF.

ly. The general form of the voltage dependence of the lasing intensity in the 1+ band system does not change appreciably in this case, i.e., the role of ultraviolet lasing must not be too great.

Another question comes to mind. Under certain conditions (particularly at pressures greater than 1.5 mm Hg and in wide tubes) lasing at a low repetition rate is observed primarily near the discharge tube walls. If the dependence of W_{pk}^{max} on the pressure and repetition rate is measured under these conditions, the results will depart rather strongly from those obtained under "unadulterated" conditions, such that the stimulated emission fills up the entire cross section of the tube. Thus, instead of a roughly linear pressure dependence of the laser power W_{pk}^{max}, this curve is seen to bend with increasing pressure (dashed curve in Fig. 13), and instead of a nearly horizontal curve for the peak power as a function of the repetition rate (Fig. 18, curve 1), the same curve exhibits a sharp ascent with increasing repetition rate (curve 2). The latter fact has a simple explanation: At a certain repetition rate the emission begins to fill up the entire cross section of the discharge tube, whereupon the laser power abruptly increases. In the more "unadulterated" case and at pressures below 1.5 mm Hg the luminescence of the discharge and stimulated emission are uniformly distributed over the entire bore of the tube, even in the one-shot triggering mode. This fact is one of the reasons that the absolute populations, gain, and other of the most important parameters were measured at the relatively low pressure of about 0.75 mm Hg.

CHAPTER V

STIMULATED EMISSION SPECTRUM OF THE $N_2(1+)$ LASER AND INTERPRETATION OF THE VIBRATIONAL–ROTATIONAL STRUCTURAL FEATURES OF THIS SPECTRUM

§ 1. The Stimulated Emission Spectrum

Pulsed lasing has been observed simultaneously at about 120 distinct lines [11, 18]. The results of the spectral measurements are given in Table 2. The measured wavelengths of the stimulated emission lines and the wave numbers calculated from them and reduced to vacuum are compared in the table with the spontaneous emission wave numbers for N_2, which are known from [56, 57]. The table gives our measured relative intensities of the emission lines, along with the rotational line to which the emission line belongs and, finally, the upper and lower laser levels. The error in the relative intensity measurement was limited chiefly by the discharge instability and amounted to about 10% for the most powerful stimulated emission lines. The discrepancy between the wave numbers for the emission lines measured on the basis of the neon lines and the data of [56, 57] does not exceed 0.08 cm^{-1}. For emission lines identified against firmly established emission lines this discrepancy is no more than 0.02 cm^{-1}. The

TABLE 2. Stimulated Emission Spectrum of N_2 Molecules in the 1+ Band System (Electron Transition $B^3\Pi_g \rightarrow A^3\Sigma_u^+$)

λ_{meas}, Å (air)	ν_{meas}, cm^{-1} (vacuum)	ν_{tab}, cm^{-1} (vacuum)	Relative intensity[a]	Rotational line	Upper and lower rotational levels[b]
		(1, 0) band[c]			
8911.27	11218.67	11218.63	40	$P_{12}3$	$F_1 3 - F_2 3$
8911.08	11218.99	11218.99		$P_{12}4$	$F_1 4 - F_2 4$
8910.63	11219.47	11219.47	210	$P_{12}5$	$F_1 5 - F_2 5$
8910.11	11220.13	11220.08		$P_{12}6$	$F_1 6 - F_2 6$
8909.55	11220.83	11220.83	330	$P_{12}7$	$F_1 7 - F_2 7$
8908.89	11221.67	11221.65		$P_1 9$	$F_1 9 - F_1 10$
8908.93	11222.87	11222.86	60	$P_{12}9$	$F_1 9 - F_2 9$
8906.63	11224.51	11224.46	20	$P_1 11$	$F_1 11 - F_2 12$
8905.66	11225.73	11225.68		$P_{12}11$	$F_1 11 - F_2 11$
8904.44	11227.27	11227.27	50	$Q_1 3$	$F_1' 4 - F_1 4$
8903.68	11228.23	11228.17		$P_1 13$	$F_1 13 - F_1 14$
8901.69	11230.74	11230.66		$Q_1 4$	$F_1' 5 - F_1 5$
8899.94	11232.95	11232.99		$P_1 15$	$F_1 15 - F_1 16$
8898.95	11234.20	11234.19	500	$Q_1 5$	$F_1' 6 - F_1 6$
8895.97	11237.96	11237.90	60	$Q_1 6$	$F_1' 7 - F_1 7$
8892.97	11241.75	11241.74	880	$Q_1 7$	$F_1' 8 - F_1 8$
8892.17	11242.76	11242.76		$P_{23}9$	$F_2 8 - F_3 8$
8891.16	11244.04	11244.06	40	$P_{23}7$	$F_2 6 - F_3 6$
8890.99	11244.26	11244.25		$P_{23}11$	$F_2 10 - F_3 10$
8890.26	11245.18	11245.18		$P_{23}6$	$F_2 5 - F_3 5$
8889.70	11245.88	11245.81	80	$Q_1 8$	$F_1' 9 - F_1 9$
8889.15	11246.58	11246.50	80	$P_{23}5$	$F_2 4 - F_3 4$
8886.41	11250.05	11250.07	1000	$Q_1 9$	$F_1' 10 - F_1 10$
8882.83	11254.58	11254.53	80	$Q_1 10$	$F_1' 11 - F_1 11$
8879.14	11259.26	11259.21	900	$Q_1 11$	$F_1' 12 - F_1 12$
8875.30	11264.13	11264.12	60	$Q_1 12$	$F_1' 13 - F_1 13$
8871.24	11269.28	11269.29	400	$Q_1 13$	$F_1' 14 - F_1 14$
8867.98	11273.43	11273.41	20	$Q_2 9$	$F_2' 9 - F_2 9$
8866.97	11274.71	11274.69	5	$Q_1 14$	$F_1' 15 - F_1 15$
8862.79	11280.03	11280.02	100[d]	$Q_2 11$	$F_2' 11 - F_2 11$
8862.52	11280.38	11280.37	100[d]	$Q_1 15$	$F_1' 16 - F_1 16$
8859.77	11283.88	11283.84		$Q_2 12$	$F_2' 12 - F_2 12$
8858.77	11285.15	11285.14		$P_1 9$	$F_1 11 - F_1 10$
8858.47	11285.53	11285.53	40	$Q_3 7$	$F_3' 6 - F_3 6$
8857.86	11286.31	11286.30		$Q_{32}5$	$F_3' 4 - F_2 5$
8857.56	11286.69	11286.68		$Q_3 8$	$F_3' 7 - F_3 7$
8856.53	11288.00	11287.99	130	$Q_2 13$	$F_2' 13 - F_2 13$
8856.28	11288.33	11288.33	150	$Q_3 9$	$F_3' 8 - F_3 8$
8854.63	11290.43	11290.43	10	$Q_3 10$	$F_3' 9 - F_3 9$
8852.99	11292.51	11292.50		{ $Q_1 17$ { $Q_2 14$	$F_1' 18 - F_1 18$ $F_2' 14 - F_2 14$
8852.61	11293.01	11292.99	200	$Q_3 11$	$F_3' 10 - F_3 10$
8850.24	11296.03	11295.99		$Q_3 12$	$F_3' 11 - F_3 11$
8849.20	11297.37	11297.33	15	$Q_2 15$	$F_2' 15 - F_2 15$
8847.58	11299.42	11299.40	140	$Q_3 13$	$F_3' 12 - F_3 12$
8844.55	11303.29	11303.21		$Q_3 14$	$F_3' 13 - F_3 13$
8844.18	11303.77	11303.74	130	$R_{21}9$	$F_2 10 - F_1 10$
8841.27	11307.49	11307.46	40	$Q_3 15$	$F_3' 14 - F_3 14$
8833.75	11317.12	11317.06		$S_{31}4$	$F_3' 5 - F_3 5$

TABLE 2 (continued)

λ meas, Å (air)	νmeas, cm⁻¹ (vacuum)	νtab, cm⁻¹ (vacuum)	Relative intensity[a]	Rotational line	Upper and lower rotational levels[b]
		(2, 1) band[e]			
8722.19	11461.86	11461.82	80	$P_{12}1$	$F_11 - F_21$
8721.99	11462.13	11462.12		$P_{12}3$	$F_13 - F_23$
8721.32	11463.01	11462.98	160[d]	$P_{12}5$	$F_15 - F_25$
8720.23	11464.45	11464.43	160[d]	$P_{12}7$	$F_17 - F_27$
8719.56	11465.32	11465.33	40	P_19	$F_19 - F_110$
8718.68	11466.48	11466.52		$P_{12}9$	$F_19 - F_29$
8713.37	11468.20	11468.18	40	P_111	$F_111 - F_112$
8716.45	11469.41	11469.41		$P_{12}11$	$F_111 - F_211$
8715.51	11470.62	11470.64	40	Q_13	$F_1'4 - F_14$
8714.50	11471.98	11471.97		P_113	$F_113 - F_114$
8712.95	11474.02	11474.01		Q_14	$F_1'5 - F_15$
8710.28	11477.54	11477.54	500	Q_15	$F_1'6 - F_16$
8707.49	11481.21	11481.25	20	Q_16	$F_1'7 - F_17$
8704.57	11485.07	11485.10	830	Q_17	$F_1'8 - F_18$
8702.55	11487.73	11487.76	25	$P_{23}7$	$F_26 - F_36$
8701.50	11489.12	11489.15	30	Q_18	$F_1'9 - F_19$
8700.69	11490.19	11490.20	20	$P_{23}5$	$F_24 - F_34$
8698.27	11493.39	11493.39	1000	Q_19	$F_1'10 - F_110$
8694.90	11497.84	11497.86	20	Q_110	$F_1'11 - F_111$
8691.36	11502.53	11502.54	820	Q_111	$F_1'12 - F_112$
8687.65	11507.47	11507.46		Q_112	$F_1'13 - F_113$
8685.57	11510.18	11510.10	5	$Q_{23}7$	$F_2'6 - F_36$
8683.75	11512.60	11512.60	150	Q_113	$F_1'14 - F_114$
8682.83	11513.82	11513.83		P_17 $Q_{12}13$	$F_19 - F_18$ $F_1'14 - F_213$
8675.57	11523.46	11523.45	70	Q_211	$F_2'11 - F_211$
8675.39	11523.70	11523.68	70	Q_115	$F_1'16 - F_116$
8671.32	11529.11	11529.08		Q_37	$F_3'6 - F_36$
8670.44	11530.27	11530.24		Q_38	$F_3'7 - F_37$
8669.55	11531.46	11531.42		Q_213	$F_2'13 - F_213$
8669.20	11531.93	11531.89	160	Q_39	$F_3'8 - F_38$
8667.62	11534.03	11534.00		Q_310	$F_3'9 - F_39$
8666.23	11535.87	11535.83		Q_117	$F_1'18 - F_118$
8665.69	11536.59	11536.54	230	Q_311	$F_3'10 - F_310$
8663.45	11539.58	11539.54		Q_312	$F_3'11 - F_311$
8662.54	11540.77	11540.74		Q_215	$F_2'15 - F_215$
8660.89	11543.03	11542.96	140	Q_313	$F_3'12 - F_312$
8658.02	11546.81	11546.80		Q_314	$F_3'13 - F_313$
8654.85	11551.04	11550.98	20	Q_315	$F_3'14 - F_314$
8653.31	11553.10	11553.10	10	$R_{21}10$	$F_211 - F_111$
		(2, 0) band			
7743.83	12909.96	12909.87	640	Q_15 P_117	$F_1'6 - F_16$ $F_117 - F_118$
7712.04	12963.17	12963.14	1000	Q_117 Q_39	$F_1'18 - F_118$ $F_3'8 - F_38$

TABLE 2 (concluded)

λ_{meas}, Å (air)	ν_{meas}, cm^{-1} (vacuum)	ν_{tab}, cm^{-1} (vacuum)	Relative intensity[a]	Rotational line	Upper and lower rotational levels[b]
		(3, 1) band[f]			
7625.67	13109.99	13109.93	8	$P_{12}5$	$F_1 5$ — $F_2 5$
7625.11	13110.96	13110.95	12	$P_{12}7$	$F_1 7$ — $F_2 7$
7622.96	13114.66	13114.69	30	$P_{12}11$	$F_1 11$ — $F_2 11$
7621 *	13117 *		5		
7617.36	13124.31	13124.29	40	$Q_1 5$	$F_1'6$ — $F_1 6$
7613.25	13131.35	13131.36	670	$Q_1 7$	$F_1'8$ — $F_1 8$
7609.85	13137.25	13137.23	30	$P_{23}5$	$F_2 4$ — $F_3 4$
7608.81	13139.04	13139.02	1000	$Q_1 9$	$F_1'10$ — $F_1 10$
7603.97	13147.42	13147.38	65	$Q_1 11$	$F_1'12$ — $F_1 12$
7598.67	13156.58	13156.52	25	$Q_1 13$	$F_1'14$ — $F_1 14$
7593 *	13166 *		2		
7592 *	13168 *		3		
7589.86	13171.86	13171.88	30	$Q_2 12$	$F_2'12$ — $F_2 12$
7587 *	13175 *		60		
7586.45	13177.73	13177.76	25	$Q_3 9$	$F_3'8$ — $F_3 8$
7584.19	13181.70	13181.66	20	$Q_3 11$	$F_3'10$ — $F_3 10$
7581.05	13187.16	13187.17	1000	$Q_3 13$	$F_3'12$ — $F_3 12$
7572.23	13202.53		20		
		(4, 2) band[c]			
7482 *	13364 *	13369.59	170	$Q_1 13$	$F_1'14$ — $F_1 14$
7487 *	13355 *	13360.42	1000	$Q_1 11$	$F_1'12$ — $F_1 12$
7491 *	13350 *	13356.14	80	$Q_1 10$	$F_1'11$ — $F_1 11$
7493 *	13347 *	13352.06	900	$Q_1 9$	$F_1'10$ — $F_1 10$
7496 *	13340 *	13344.41	100	$Q_1 7$	$F_1'8$ — $F_1 8$
		(0, 0) band			
10477 *	9545 *	9646.68	60	$Q_1 9$	$F_1'10$ — $F_1 10$
10488 *	9538 *	9539.76	1000	$P_{23}7$	$F_2 6$ — $F_3 6$
10496 *	9527 *	9529.67	65	$Q_1 5$	$F_1'6$ — $F_1 6$
10498 *	9525 *	9526.04	430	$P_1 13$	$F_1 13$ — $F_1 14$
10504 *	9515 *	9516.82	130	$P_{12}7$	$F_1 7$ — $F_2 7$

[a] Experimental conditions: p = 3.5 mm Hg; V = 25 kV; f = 8 Hz; transmissivities of silvered mirrors, 0 and 4%; discharge tube bore, 15 mm; tube length, 1.3 m; C = 0.01 μF. In each band the intensity of the strongest line is adopted as 1000.

[b] The upper (left) and lower (right) laser levels in column 6 are designated by the quantum numbers J' and J", respectively. The levels for the ortho modification of molecular nitrogen are underlined.

[c] Data of [56] for the entire band.

[d] The total intensity of these emission lines was measured by the photoelectric method. The relative intensity was determined visually from spectrograms.

[e] Data of [57] for the entire band.

[f] Computed from the level tables in [56].

starred wavelengths were measured correct to ~ 10 Å. The postulated allocation of these lines is given in the table.

The vibrational and rotational structure of the stimulated emission spectrum displays marked differences from the spontaneous emission spectrum. Of the ~ 100 bands detected in the spontaneous emission spectrum for the 1+ band system of N_2, only bands corresponding to transitions from lower vibrational levels of the upper laser state $B_{v'}$ ($v' < 4$) to lower vibrational levels of $A_{v''}$ ($v'' < 2$) of the A state are observed in stimulated emission (the state $A^3\Sigma_u^+$, which is called the A state for short, is the lower working state of the $N_2(1+)$ laser). A large part of the laser beam energy belongs to the (1, 0) and (2, 1) bands, which have roughly equal intensity. These bands have the best-developed rotational structure, 54 stimulated emission lines being detected in (1, 0), and 39 in (2, 1). The spectrum of these bands was the most assiduously investigated. The other bands are considerably weaker and have fewer lines.

The stimulated emission bands (1, 0) and (2, 1) have almost the same rotational structure. A large fraction of the laser emission lines observed in these bands belongs to the following branches: Q_1, P_{12}, Q_3, Q_2, P_{23}, and P_1 (27 branches can be observed in the spontaneous emission bands [80]). Under the conditions described in the footnotes to the table about 70% of the intensity of each of the bands (1, 0) and (2, 1) is ascribed to the Q_1 branch, in which the greatest number of lines is excited: 14. The lines belonging to the ortho modification of the nitrogen molecule carry the preponderant share of the energy of this branch. In all other branches and bands the ortho-modification lines are also considerably stronger than the para-modification lines.

The intensity distribution of the laser emission lines in the (1, 0) band is shown in Fig. 19 as a function of the total angular momentum J' in the upper laser state. As the figure indicates, the intensity maximum in different branches is attained for different values of J'. The most intense branch Q_1 plus the branch Q_3 have a maximum at J' = 10. The three strongest lines of the Q_1 branch, i.e., Q_17, Q_19, and Q_111, carry about 70% of all the energy for the branch [i.e., ~ 50% of the power for the (1, 0) band]. In the P_{23}, P_{12}, and Q_2 branches the maximum is reached for J' = 6, 7, and 12, respectively. The (2, 1) band exhibits a similar dependence. The intensity distribution among the rotational lines of the (2, 0), (3, 1), and, evidently, (0, 0) bands differs markedly from the distribution indicated in Fig. 19.

We have already mentioned in Chapter I the investigation [41], in which the stimulated emission lines have been identified independently. The total number of lines observed in [41] was somewhat smaller than in our studies [11, 18]. With rare exception, almost all the laser emission lines for which the wavelengths were measured exactly were allocated to the same

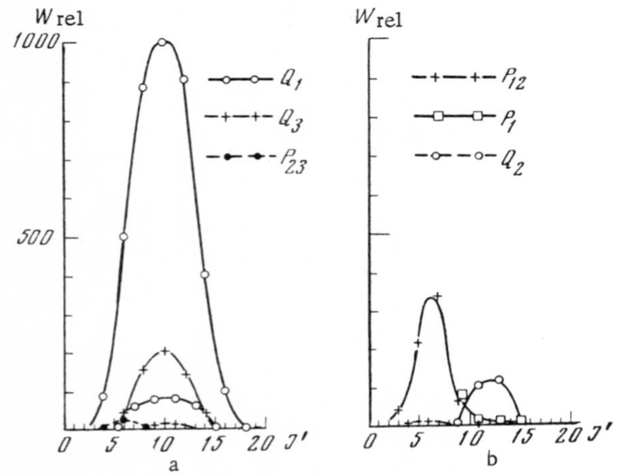

Fig. 19. Relative intensity of laser emission in the (1, 0) band versus total molecular angular momentum J' in the upper laser state. Intensity of the Q_19 line adopted as 1000; ortho modifications of molecular nitrogen belong to: a) levels with even J'; b) levels with odd J'. An analogous dependence obtains for the (2, 1) band.

molecular transitions in our work as in [41]. Several new stimulated emission lines in the 1+ band system have been discovered in [81].

§2. Electron States and Transitions in N_2; Analysis of Excitation of the Vibrational and Rotational Structure of the Laser Emission Spectrum

The potential curves for the electron states that interest us in molecular nitrogen have been computed in [82] and are presented in Fig. 20. Like the H_2 molecule potential curve, here the curve for the lower laser A state has the maximum shift relative to the ground state. Also shown in addition to the potential-energy curve for the upper laser B state is the curve for the C state, i.e., the upper working state of the $N_2(2+)$ laser, as well as the curve for the $a^1\Pi_g$ singlet. As postulated in [3], it is from the latter state that the upper working levels of the $N_2(1+)$ laser are populated via step-by-step transitions.

Transitions from the triplet A, B, and C states to the ground states have low probability, because they incur a change of spin. The radiative lifetime of the lower laser A state is so great (~ 2 sec [83, 84]) that this state can be regarded unconditionally as metastable.

The most intense triplet band systems in the spontaneous emission of N_2 are the first positive and second positive systems, corresponding to B–A and C–B transitions. The 2+ band system is observed approximately in the interval from 3000 to 5000 Å, and the 1+ system from 5000 Å and well into the infrared region of the spectrum. The oscillator strength f_e for the latter system is known from [85, 86] with an error of ~ 30%.

The radiative lifetime $\tau_{B_{v'J'}}$ of the individual rotational level J' of the upper laser state, $B_{v'}$, is determined by the total probability of transition to levels of the A state:

$$\tau_{B_{v'J'}} = \Big(\sum_{v''}\sum_{J''} A^{B_{v'J'}}_{A_{v''J''}}\Big)^{-1} = \left(\sum_{v''}\sum_{J''} \frac{64\pi^4 (\nu^{B_{v'J'}}_{A_{v''J''}})^3}{3hc^3} \frac{S^{BA}_e}{g'_e} q_{v'v''} \frac{S_{J'J''}}{2J'+1}\right)^{-1}, \tag{7}$$

where $A^{B_{v'J'}}_{A_{v''J''}}$ is the Einstein coefficient for the transition $B_{v'J'} \to A_{v''J''}$ [86], the frequency of which is $\nu^{B_{v'J'}}_{A_{v''J''}}$ (in Hz); S^{BA}_e is the strength of the electron B–A transition; $g'_e = 2$ is the electronic statistical weight of the B state; $q_{v'v''}$ are the Franck–Condon factors for the transition $v' \to v''$; and $S_{J'J''}$ is the Hönl–London factor.

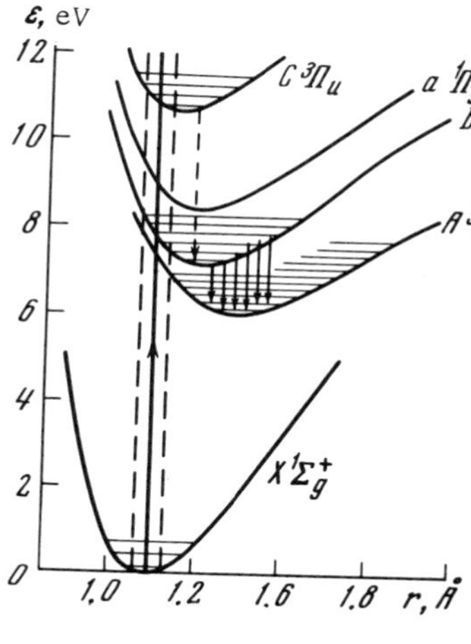

Fig. 20. Potential-energy curves for the electron states of molecular nitrogen.

Recognizing the fact that in intraband transitions the quantity $\nu_{A_{v''}J''}^{B_{v'}J'}$ changes very slightly and that $\sum_{J''} S_{J'J''} = 2J' + 1$, for $\tau_{B_{v'}J'}$, we deduce the expression

$$\tau_{B_{v'}J'} = \left(\sum_{v''} \frac{64\pi^4 (\nu_{A_{v''}}^{B_{v'}})^3}{3hc^3} \frac{S_e^{BA}}{g_e'} q_{v'v''} \right)^{-1} = \tau_{B_{v'}}. \tag{8}$$

As implied by (8), different rotational levels ascribed to a given vibrational state have the same radiative lifetime $\tau_{B_{v'}}$. The value of $\tau_{B_{v'}}$ is easily calculated from the known [87] Franck–Condon factors and the known [56] frequencies $\nu_{A_{v''}}^{B_{v'}}$. For the v' = 0, 1, 2, 3 levels the radiative lifetime $\tau_{B_{v'}}$ turns out to be equal, respectively, to 14, 10, 7, 5, and 6 μsec. The radiative lifetime of levels of the C state is approximately two orders of magnitude smaller, about 50 nsec [88].

In the analysis of the population of the vibrational levels, as in the case of the H_2 laser, the influence of vibrational relaxation can be neglected [74, 75]. As for the Franck–Condon principle, we know [65] that it is even better satisfied for heavy nuclei than for light ones (the N_2 nucleus is seven times heavier than that of H_2).

At the beginning of pulsed discharge the lower vibrational levels of the B state (v' = 0, 1, 2, 3, 4; see Fig. 20) are the most abundantly populated, by "direct electron impact." The v'' = 0, 1, 2 levels of the lower laser state remain "empty" in this case, because their excitation by electrons from the ground state violates the Franck–Condon principle. Lasing is also possible between these vibrational levels. The corresponding transitions are designated by vertical strokes in Fig. 20, and they are the ones reponsible for stimulated emission. As we see, the situation is largely similar to that in the case of hydrogen. As in the case of the H_2 laser, the "direct electron impact" mechanism helps us at once to understand why, of the many bands that can be observed in spontaneous emission in the 1+ band system of N_2, only a few bands, and namely the ones just described, are observed in laser emission.

The analysis of the rotational structure of the stimulated emission spectrum has certain singularities. The rotational-level and transition diagram is identical on the whole for different bands of the 1+ system. It is shown for the (1, 0) band in Fig. 21.

The lower laser state consists of three groups of rotational levels, F_1, F_2, and F_3, with numbers Ω = 0, 1, and 2, respectively (Ω is the total electron angular momentum of the molecule). Levels having equal K (K is the total angular momentum of the molecule, minus spin) do not differ in energy by more than 2 or 3 cm^{-1}. The shifts between the groups of levels $^3\Pi_0$, $^3\Pi_1$,

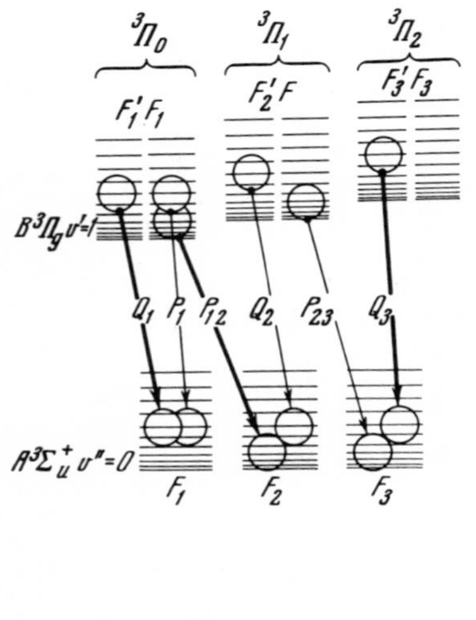

Fig. 21. Laser-level diagram for the (1, 0) band [56]. The branches observed in laser emission are shown.

and $^3\Pi_2$ of the upper laser B_1 state are much greater, amounting to ~ 50 cm^{-1}. Each rotational level of the latter groups is split into two half-levels of practically equal energy. Spontaneous transitions from all rotational levels of the B_1 state to all levels of the A_0 state then form the (1, 0) band, in which upwards of several hundred lines are observed. The rotational lines are grouped by branches (a total of 27 branches) according to the change in total angular momentum J of the molecule, as well as according to the initial and final groups of levels. Our notation convention is the same as in [56, 65], except in one case: We do not use the quantum number K as in [56], but the total momentum J to enumerate the levels of the A_0 state. For the enumeration of the spin components of the lower laser state, as in [56], we use subscripts 1, 2, and 3, respectively, to denote the additions of momenta J = K + 1, J = K, and J = K − 1. The subscripts 1, 2, and 3 for the upper laser state exceed the number Ω by one unit. The rotational lines are designated by the number K for the lower level.

We now analyze the excitation of the rotational structure of the laser emission spectrum on the basis of a simplified model. As shown in [77], in problems of this type it is useful to compute the relative gain for the rotational lines and then, using the qualitative relationship between the gain and emission intensity, to draw a comparison with the experimental. We assume in estimating the gain that the population distribution among the rotational laser levels is thermal. For the ground state the thermal distribution is established very quickly, in about five collisions [74, 75]. The same number of collisions is suffered by ground-state nitrogen molecules about $2 \cdot 10^{-7}$ sec later at a temperature of 320°K and pressure of 3.5 mm Hg. Due to the large gas-kinetic dimensions involved the thermal distribution must be established even faster for the laser states. The influence of spontaneous emission on the establishment of the thermal distribution can be neglected, because the radiative lifetime of the upper laser states is relatively large (~ 10 μsec), and the lower levels are metastable. The true population distribution among the rotational levels will be closer to the thermal model the greater the inversion holding time and the higher the gas pressure. Deviations from a thermal distribution should be observed only in the interval of very low pressures (~ 0.2 mm Hg) and for short laser emission pulses.

The gain $k_{A_{v''J''}}^{B_{v'J'}}$ in the center of the Doppler-broadened rotational line at the electronic vibrational–rotational transition $B_{v'J'} \to A_{v''J''}$ is determined by the expression

$$k_{A_{v''J''}}^{B_{v'J'}} = \frac{16(\pi^5 \ln 2)^{1/2}}{3hc} \frac{\nu}{\Delta\nu_D} \frac{S_e^{BA}}{g_e'} q_{v'v''} \frac{S_{J'J''}}{2J'+1} \left(N_{J'} - N_{J''} \frac{g_e'(2J'+1)}{2J''+1} \right), \tag{9}$$

in which ν is the transition frequency (in Hz), $\Delta\nu_D$ is the half-width of the Doppler line contour, and $N_{J'}$ and $N_{J''}$ are the populations of the upper and lower rotational levels. Equation (9) is easily derived from the conventional gain equation as given, for example, in [78] by substituting into the latter equation the corresponding expression for the Einstein coefficient of the molecular transition. Following [86], we use the electron transition strength S_e^{BA}, which is proportional to the Einstein coefficient.

Taking account of the fact that for a thermal population distribution among the rotational levels the quantity $N_J = \frac{N_v}{Z_v} g_N(2J+1) e^{-E_J/kT} g$ (N_v is the total population of the vibrational level, Z_v is the statistical sum of the vibrational state, g_N – equal to 2 or 1 – is the nuclear statistical weight, E_J is the energy of the rotational state, and T_g is the temperature of the gas), we readily obtain for the gain

$$k_{A_{v''J''}}^{B_{v'J'}} = \frac{16(\pi^5 \ln 2)^{1/2}}{3hc} \frac{\nu}{\Delta\nu_g} \frac{S_e^{BA}}{g_e'} q_{v'v''} S_{J'J''} g_N \left\{ \frac{N_{v'}}{Z_{v'}} e^{-E_{J'}/kT_g} - \frac{N_{v''}}{Z_{v''}} g_e' e^{-E_{J''}/kT_g} \right\}. \tag{10}$$

In the present section we are concerned primarily with the value of the gain at individual rotational lines within a band in the amplification mode, when the population of the lower laser state in (10) can be neglected. The latter step is valid at least for the v'' = 0, 1, 2 vibrational levels of the lower laser state, because, as mentioned above, the population of these levels by electron impact from the ground state has exceedingly low probability due to violation of the Franck–Condon principle. Under these conditions the expression for the relative intraband gain is

$$k_0(J') = \text{const}\, g_N S_{J'J''} \exp\left\{-E_{J'}\frac{hc}{kT_g}\right\}. \tag{11}$$

The quantity $k_0(J')$ is easily calculated for different branches and different values of J'. The factors $S_{J'J''}$ and the quantities $E_{J'}$ can be determined from equations given in [89] and [90], respectively. The temperature of the gas was assumed equal to 320°K in the calculations. The calculated values of $k_0(J')$ are depicted graphically in Fig. 22. As evident from the figure, the gain differs for different branches and is strongly dependent on the value of J'. The maximum gain for different branches is attained at strikingly different values of J'. For example, the maximum gain in the P_{12} branch is attained at J' = 6, whereas in the Q_2 branch it is attained at J' = 12.

The principal features of the rotational structure of the stimulated emission spectrum are now brought better into focus. For example, the considerable disparity observed experimentally in the intensities of the laser emission lines for the ortho and para modifications of N_2 (cf. Figs. 19 and 22) is meaningful. We need only consider the abrupt increase of the emission intensity as the gain is raised above the threshold value. The greatest emission intensity must in fact occur for the branch Q_1, because the gain is maximal in this case. Only a few

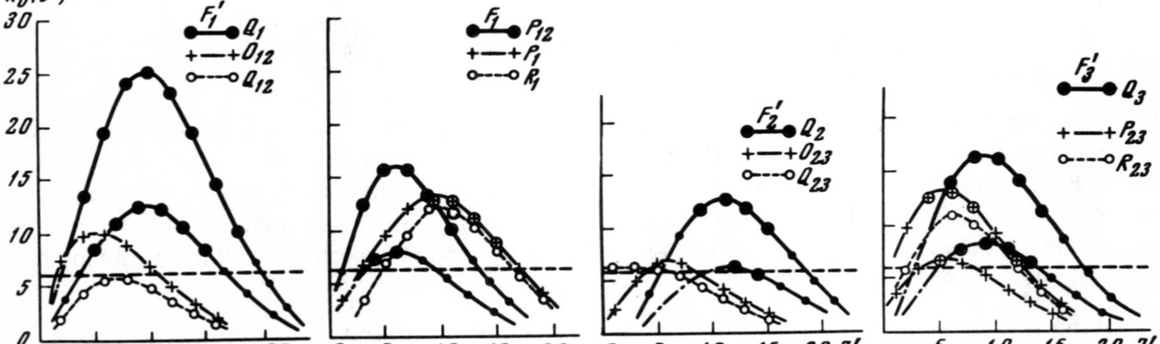

Fig. 22. Gain at rotational lines versus J' for upper laser level, calculated according to Eq. (11). The horizontal dashed line gives the threshold gain. Ortho-modification levels: a and d) even J'; b and c) odd J'. F_1', F_2', etc., correspond to rotational-level groups of the upper laser state. The larger dots and circled crosses correspond to lines observed in laser emission.

branches have gains above the threshold, clearly because only six of the 27 stimulated emission branches are observed. It is important to note that certain branches having an above-threshold gain cannot be observed in laser emission due to the "suppressant" action of stronger branches having the exact same initial levels (for example, the Q_{12} branch can be suppressed by the Q_1 branch, and the R_1 branch by the P_1 branch).

Comparing Figs. 19 and 22, we are readily convinced that for all the branches observed in laser emission expression (11) in general gives the true value of J' corresponding to the strongest emission lines within a branch. For example, in the P_{12} branch the emission maximum is attained for J' = 6, and in the Q_2 branch for J' = 12, as is fully consistent with the calculations. The identity of the rotational structure of the stimulated emission spectrum in the (1, 0) and (2, 1) bands also follows from Eq. (11), insofar as the Hönl–London factors and the quantities $E_{J'J''}$ for the (2, 1) band are practically the same as for the (1, 0) band.

We see, then, that the experimental facts can be semiqualitatively accounted for solely on the basis of the relatively straightforward approximations adopted above in the gain calculation. The satisfactory agreement of the analytical and measured values indicates, in particular, that the thermal population distribution among the rotational states is, for gain calculations at any rate, a good approximation to the true distribution.*

Proceeding from the thermal population distribution among the rotational levels (this hypothesis, in light of the foregoing discussion, should be reasonably indisputable) and taking account of certain experimental data, we can now show with little difficulty that the approximations used in the derivation of Eq. (11) must be very close to reality. One of these approximations is the neglect of the population of the lower laser state. Regarding the relative population of the lower and upper vibrational states $N_{v''v'} = \delta$, we can assess this ratio on the basis of the relative laser emission intensity at the rotational lines belonging to different branches. As a matter of fact, the relative gain and, hence, intensity of the laser emission in certain branches depend very strongly on δ. This situation is physically related to the different population distributions among the groups of levels F_1, F_2, and F_3 for the upper and lower laser states. The most appreciable variations with respect to the gain must be observed in the branches with upper levels F_3 and F_1. Proceeding from the general relation (10), we readily calculate the corresponding relative gain, the expression for which contains, as a parameter, the relative population of the upper and lower vibrational states. Calculations were carried out for the relative gain specifically at the $P_{12}5$ and $Q_3 11$ lines of the (1, 0) band. The laser emission intensities at these lines are approximately equal (see Table 2), attesting to an approximately equal gain. For the relative gain k_3/k_{12} at the $P_{12}5$ and $Q_3 11$ lines we obtain

$$\frac{k_3}{k_{12}} = 1.2 \cdot \frac{1 - 4\delta}{1 - 2.9\delta} \cdot \qquad (12)$$

(The calculations were carried out for T_g = 320°K.) The equality of the emission intensities at these lines implies that $k_3/k_{12} = 1$. We then obtain from (12) $\delta \cong 0.2$, i.e., consistent with the foregoing assumptions, the population of the lower vibrational state is one fifth that of the upper state.

Another significant assumption underlying the analysis of the rotational structure of the stimulated emission spectrum is contained in the fact that the temperature of the gas is considered to be roughly equal to room temperature, 320°K. This assumption is readily verified

*A largely analogous treatment of the excitation of the rotational structure of the stimulated emission spectrum in the 1+ band system of N_2 has been carried out in the previously cited paper [41], which was published a short time after our paper [11]. The relative dependences $k_0(J')$ obtained in [41] concur with our calculations in [11, 18] to within 10%.

according to the relative intensity of the emission lines of the Q_3 branch. The latter has the singular feature that the exponents entering into (10) for the upper and lower levels agree with a high degree of proximity. The expression for the relative gain k_J at the $Q_3(J)$ lines has the form

$$k_J = \text{const}\, S_J \exp\left\{-J(J+1)\frac{1{,}73}{kT}\right\}. \tag{13}$$

The factor S_J for the Q_3 branch can be fairly well approximated by the simple expression $S_J \cong J - 3$. From the maximum condition for (13) we readily deduce the following relation between the value of J corresponding to maximum gain (stimulated emission) and the temperature T_g:

$$(J-3)(2J+1)\frac{2{,}45}{T_g} = 1. \tag{14}$$

Inasmuch as the maximum emission intensity in the Q_3 branch is attained for J' = 10 (see Fig. 19), we infer from (14) that T_g = 370°K. This corroborates the assumption of a relatively low gas temperature.

We now take a brief look at some still faintly obscure problems. The solution of these problems is not of prime importance, because they merely pertain to the details of the lasing mechanism. Within the scope of the above-analyzed inversion mechanism between the vibrational levels it is a little difficult to understand at first glance why, although the laser power in the (1, 0) and (2, 1) bands is very large, only very weak laser emission is observed in the (2, 0) band. The probability of vibrational transition in the (2, 0) band is practically the same as in the (2, 1) band, but decay takes place at the A_0 level, whose filling from the ground state must be even weaker than that of A_1, which is the lower laser level for the (2, 1) band. It would seem, therefore, that the laser emission in the (2, 0) and (2, 1) bands should have approximately the same intensity, whereas in fact the laser power in the former band is considerably weaker than in the latter. Clearly, the explanation of this dilemma must be sought in the possibility of "competition" processes, similar to the previously cited "competition" of the P and R branches in the hydrogen laser, between different laser emission bands having congruent upper or lower working levels.

Thus, as Fig. 23 discloses, the (2, 0) band has the upper v' = 2 level in common with the (2, 1) band and a lower level in common with the (1, 0) band. In the (1, 0) band, which has twice the transition probability, lasing develops more rapidly than in the (2, 0) band, so that the inversion between the v' = 2 and v'' = 0 levels deteriorates. Laser emission in the (2, 1) band, on the other hand, diminishes this inversion even further, because the molecules in this case are ejected from the v' = 2 to the v'' = 1 level. The culminative result is the inhibition of laser action in the (2, 0) band. A similar type of "competition" obviously obtains between the (0, 0) and (0, 1) bands, the (0, 0) and (1, 0), and the (2, 1) and (3, 1) bands.

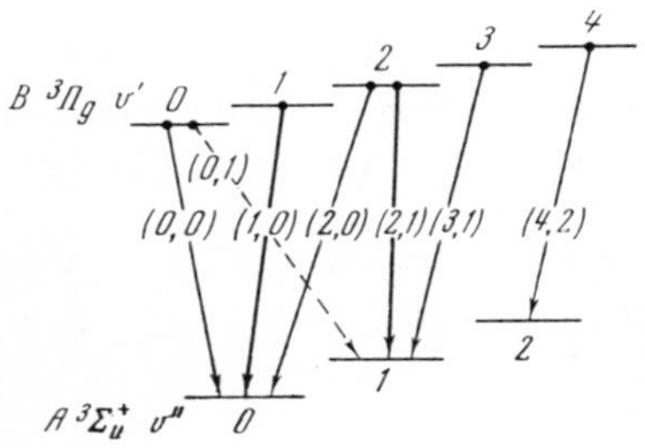

Fig. 23. Vibrational-level diagram of the upper and lower working states of an $N_2(1+)$ laser. The transitions observed in laser emission are marked with arrows. The emission intensity is conditionally indicated by the thickness of the line.

Processes of the type described above also have a powerful influence on the rotational structure of the laser emission spectrum. Highly characteristic is the fact that, by contrast with the (1, 0) and (2, 1) bands, whose laser emission occurs simultaneously at several dozen lines, in the (2, 0) band under any excitation conditions laser emission is observed at two somewhat resolved lines, and we note that the analogous lines in the (1, 0) and (2, 1) bands are by no means the most intense. The laser emission intensity distribution among the rotational lines of the (3, 1) and (0, 0) bands is also different from the distribution that would be expected on the basis of Eq. (11). Not to be overlooked is the possibility that these disparities, like the singular attributes of the rotational structure of the (2, 0) band, would be better understood with allowance for the perturbing action of powerful laser emission in the (1, 0) and (2, 1) bands.

CHAPTER VI

ASPECTS OF THE PHYSICS OF PULSED DISCHARGE

§1. Pulsed-Discharge and Inversion Dynamics of the $N_2(1+)$ Laser

The conditions present in pulsed gas lasers are such as engender diverse gas-discharge processes associated with breakdown of the working gap. As a rule, the breakdown in this case occurs in a long cylindrical tube at a low gas pressure (~ 1 mm Hg) in a strongly inhomogeneous field and at relatively high overvoltages. Under these conditions potential and luminescence waves have often been observed propagating at a high speed along the discharge tube, with the formation of a longitudinal field [91-93], under conditions of a rapidly increasing voltage [94], and in transition from glow discharge to arc discharge [95]. In the initial phase of discharge electron-optical effects have also been noted [96, 97]. In [91-97] essentially the same process is treated from different standpoints, namely the process of the development of discharge in long tubes. The investigations of several authors indicate that the formation of pulsed discharge must be divided into separate stages [98-100]. The general pattern of development of pulsed discharge under the conditions named above, however, is not clear at the present time.

All of these considerations greatly impede the analysis of the operation of pulsed gas-discharge lasers. It was particularly urgent for us to answer the question of the stages run through by the pulsed discharge used in lasers, of which stage of discharge contributes the most to the formation of inversion, and of the physical conditions in the discharge at the time of laser emission. The detailed investigation of all stages, omitting none, of pulsed discharge is purely a gas-discharge problem. We are concerned almost exclusively with problems whose solution is prerequisite to understanding the dynamics of the physical processes involved in pulsed gas lasers. In the present chapter we concentrate on these problems in the greatest detail. The results are given primarily for the $N_2(1+)$ laser. To facilitate the presentation we analyze the experimental data separately for each stage of discharge.

The general pattern of the discharge developmental process shows up the most clearly and completely in long tubes at low pressures, on thoriated tungsten cathodes, and at small overvoltages [101]. The discharge dynamics for positive and negative voltages is identical in general, differing only in the details (Fig. 24). The experiments have been carried out, as a rule, with a positive voltage. The time dependences given in Fig. 24 were obtained by the processing of many oscillograms, examples of which are shown in Figs. 25 and 26.

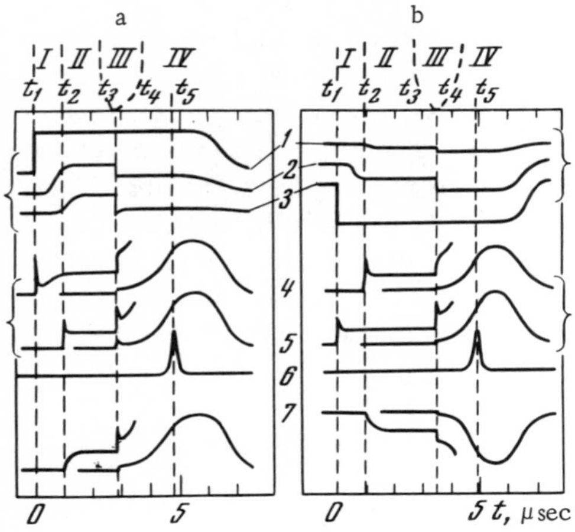

Fig. 24. Typical oscillograms of positive (a) and negative (b) voltages. 1-3) Potential in the vicinity of the anode, in the midsection of the tube, and in the vicinity of the cathode; 4, 5) spontaneous emission in the anode and cathode regions; 6) stimulated emission; 7) current. The initial parts of oscillograms 4, 5, and 7 are given on a blown-up scale. The time scale and relative durations of the stages are to be considered as approximate. Typical conditions: p, 0.5 mm Hg; V = 15 kV; l = 6 m; C = 0.15 μF; current, ~1.5 kA; thoriated tungsten electrodes.

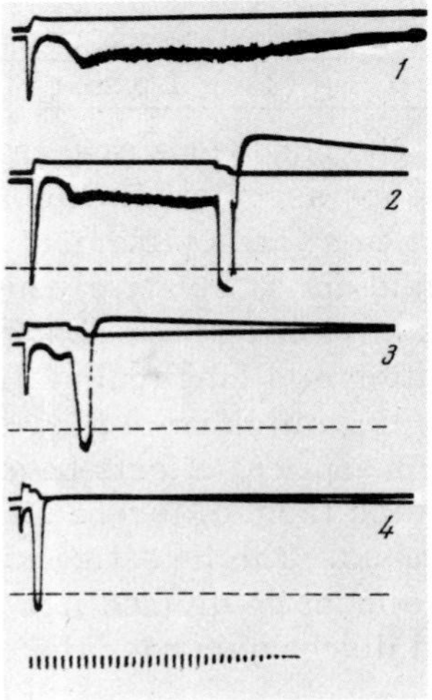

Fig. 25. Development of pulsed discharge with increasing voltage. 1-4) Oscillograms of voltage in tube midsection (upper of each pair of curves) and spontaneous emission in anode region at voltages of 10, 10.4, 10.8, and 12.5 kV, respectively. Time marker, 10 μsec; p = 0.7 mm Hg; l = 3.5 m; C = 0.04 μF. The dashed horizontal line, both here and in Fig. 26, indicates the linear-deflection limit of the oscilloscope beam.

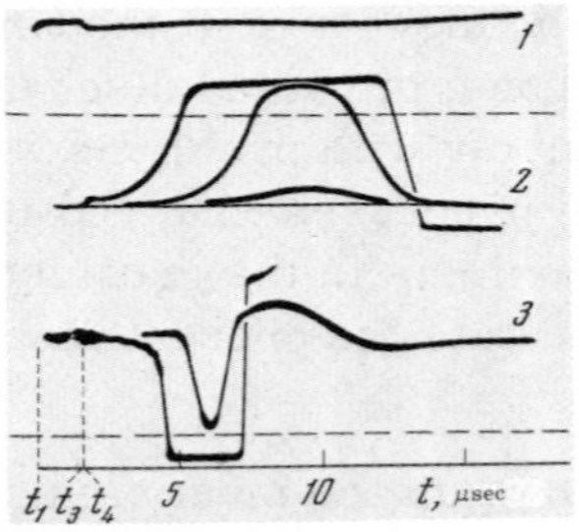

Fig. 26. Development of current and stimulated emission in the 1+ band system of N_2. 1) Potential in tube midsection; 2) current; 3) stimulated emission; V = 13.6 kV; p = 0.65 mm Hg; l = 3.5 m; C = 0.15 μF; current, 1.2 kA; thoriated tungsten cathode. The current oscillograms are shown in 1/10 and 1/100 attenuation. The attenuation of the stimulated emission pulse is arbitrary.

We can segregate the development of pulsed discharge into four distinct phases, in which the physical conditions are markedly dissimilar. The conditions most favorable for the formation of inversion occur in the last, high-current phase of discharge. This phase is treated in the greatest detail. The development of the physical processes in the first three stages is largely of secondary interest where our problem is concerned and is given merely to complete the total picture.

First Stage ($t_1 < t < t_2$): Longitudinal Field Formation. In view of the fact that discharge tubes are mounted near a grounded metal baseplate the electrostatic field in the absence of conduction currents is concentrated mostly near the cathode and anode and therefore does not exert any appreciable influence on the formation of conduction over the entire length of the tube. Under these conditions, as shown in [91-93, 98] and as observed in our own work, a potential wave is generated in the discharge tube, where it is accompanied by luminescence and propagates from the anode to the cathode for a positive voltage. The speed of this wave increases very significantly with the voltage. Even at relatively low voltages (~10 kV, p = 0.2 mm Hg) the propagation speed of the wave is about $2 \cdot 10^8$ cm/sec. At time t_2 the luminescence wave reaches the cathode, and the tube acquires a small conductivity over its entire length.

Second State ($t_2 < t < t_3$): Dense Glow Discharge. In this stage luminescence is observed over the entire length of the discharge tube, sometimes acquiring the typical striated quality of glow discharge. A cathode spot is not formed, so that the discharge does not gain an arc character in the second stage. The current density does not exceed a few amperes per square centimeter. The potential distribution over the length of the tube is nonuniform, the field being concentrated predominantly near the cathode (Fig. 27), indicating the existence of a well-developed positive charge in the cathode region. This stage is clearly a modification of dense glow discharge [102, 103].

Third Stage ($t_3 < t < t_4$): Potential Wave Formation. The processes in this phase of discharge evolve very rapidly, even at low overvoltages and for a discharge gap length of ~6 m. At the time t_3 a potential wave is initiated in the vicinity of the cathode and propagates along the tube toward the anode. Within the error limits of our measurements the onset of the current pulse and decay of the potential near the cathode (see Fig. 24) occur simultaneously. The current pulse width is apparently shorter than 100 nsec, and the amplitude depends strongly on many conditions, varying from fractions of an ampere to tens of amperes. After the cathode region the potential decay propagates at a high speed (~$5 \cdot 10^9$ cm/sec) along the tube toward the anode. As a result of the transmission of this wave along the tube the potential distribution along the tube becomes increasingly uniform, and the degree of ionization and current jump abruptly.

Processes of this nature have been observed earlier under similar conditions [95] in a study of the transition from glow to arc discharge. According to [95], potential waves are generated by the virtually instantaneous (5 nsec) increase by many orders of magnitude of the cathode emissivity associated with the formation of the cathode spot and the concomitant sharp increase in the number of electrons in the cathode sheath. The propagation of the potential wave in the tube induces large local gradients, which move with a speed of 10^9 cm/sec and are accompanied by a corresponding ionization and photoemission.

The physical picture of the processes on which these phenomena are based is still obscure. It can only be hypothesized that the "electron-shock-wave" model is valid in this case [104].

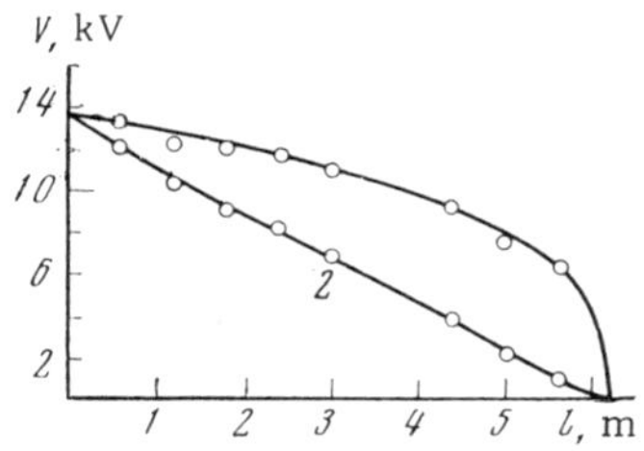

Fig. 27. Potential distribution along tube in the glow and arc discharge stages for a positive voltage. 1) Glow discharge stage; 2) arc discharge stage; p = 0.2 mm Hg.

Fourth Stage ($t > t_4$): Pulsed Arc Discharge. This stage in the development of discharge, particularly in its incipient phase, is characterized by a relatively slight time variation of the potential and a rapid increase in the current and emission intensity. The field can be thought of as approximately uniform (see Fig. 27). The uniformity of the potential distribution attests to the absence of space charges of any appreciable magnitude. The rapid growth of the current gives rise to intense filling of the molecular levels. We especially emphasize the fact that stimulated emission occurs immediately after the formation of the cathode spot, when the current growth is no longer limited by inadequate electron emission from the cathode surface and the role of cathodic processes in the current growth is, for the most part, minor.

The emission of electrons by the cathode, as we know (see, e.g., [105]), is sometimes attended by "plumes," i.e., ejections of matter from the cathode at a speed of 10^6 cm/sec (this happens when the cathode spot functions in the thermal mode). In a period of 5 to 10 μsec the vaporized metal is displaced over relatively small distances ($\sim$10 cm) and therefore does not exert a significant influence on the discharge conditions. This inference is supported, in particular, by the fact that the discharge emission spectra do not contain lines for the electrode material.

The maximum contribution to the population of the molecular states occurs specifically in the high-current phase of discharge. The filling of the laser levels in the first through third stages, as a rule, can always be neglected. The intensity of spontaneous emission in the first three stages is one or two orders of magnitude weaker than in the high-current phase. The role of the first three stages is essentially a preparatory activity leading up to the transition to the high-current discharge phase. The initial current and initial degree of ionization in this phase are determined by the first stages of discharge in particular.

The physical mechanisms of the first three stages are not even qualitatively clear in certain respects. This situation, however, does not present fundamental difficulties in comprehending the excitation dynamics of the given laser, because at the time of inception of laser emission the plasma has largely managed to relax, and the prior history of the discharge has become for the most part inconsequential.

With an increase in voltage the general pattern of discharge development remains, on the whole, the same as at low voltages, only the durations of the phases, particularly the first and second, being foreshortened (see Fig. 25). The delay of the stimulated emission pulse relative to the voltage leading edge is also diminished. This delay is equal to the sum of the durations of the first discharge phases and the time interval t_{max} measured from the inception of the high-current stage to the laser emission pulse. Only the second member of this sum can be interpreted quantitatively; it is associated with the current growth rate. The difficulties in the interpretation of the durations of the first two discharge phases stem from the almost total uncertainty regarding the physical processes underlying the formation of the longitudinal field and generation of the cathode spot (according to [106], the formation of the cathode spot is strongly dependent on the surface state of the cathode).

The material of the cathode has a strong influence on the breakdown behavior. Whereas the currents in the second stage are relatively low for a tungsten cathode, so that the population of the working stages in this phase can be neglected, these currents can become very large in the case of an aluminum or brass cathode. In this event it is necessary to take account of the population of the levels already in the second stage. It must be borne in mind that the current growth rate in this case is determined to a considerable extent by the γ-processes at the cathode. The allowance for these processes, which depend on the specific state of the cathode surface, is very difficult to realize. It is for this reason that the bulk of the measurements were performed with tungsten cathodes.

With the pressure elevated above 1.5 mm Hg or under not too explicit experimental conditions the pulsed discharge has a filamentation tendency in the high-current phase. This tendency is particularly strong at low overvoltages, low (1 or 2 Hz) repetition rate, and low pressures of 5 to 7 mm Hg. With a gradual increase in the gas pressure the discharge at first becomes localized near the tube walls, ostensibly forming a hollow cylinder. The luminescence of molecular nitrogen is weak in the center of the tube. Laser emission is observed in a ring configuration. With an increase in the repetition rate or overvoltage the luminescence and discharge again become homogeneous. As the pressure is further increased the discharge is drawn out into a filament, whose diameter at 10 mm Hg can become roughly equal to 0.5 to 1 cm. A characteristic kink is observed simultaneously on the current oscillogram. The filament adheres to the lower wall of the tube and has a corkscrew shape. As a rule, the kinks tend to repeat in each current pulse. With an increase in voltage the filament at first dissipates near the walls, and with an increase in repetition rate it can spread to fill up the entire cross section of the discharge tube. We note that the filamentation tendency, all things considered, is strongly affected by the state of the cathode or the impurity state of the N_2. The interpretation of the observed effects is rendered extremely difficult. It can only be hypothesized that with an increase in pressure we run up against the case of transition to the streamer discharge state. This case has been very meagerly investigated to date in the physics of breakdown.

It must also be pointed out that, in principle at least, lasing can occur in the first through the third stages, but it is of a relatively minor magnitude. As shown by repeated studies, gas-discharge processes of the "potential wave" type do not lead (as opposed to the arc phase) to stimulated emission in the given lasers. This situation is clearly attributable to the local quality of these waves, since it is clear that the gain in a homogeneous elongated plasma, all other conditions being equal, is greater than in one of bounded length. It must be realized, however, that the growth rate of the electron density in "potential waves" can be very high, so that the formation of inversion remains at least a hypothetical possibility.

The relatively long-duration laser action that we observed in H_2 (see Chapter III) occurs in the dense glow discharge phase and represents an example of lasing in discharge stages other than the arc discharge phase. However, as already mentioned, the intensity of the stimulated emission is much smaller than in the high-current discharge phase.

§2. Temporal Evolution of the High-Current Discharge Phase; Physical Model of Discharge

Under the conditions anent a homogeneous field, which, as shown in the preceding section, is realized in the high-current discharge phase and for not too high a degree of ionization of the discharge (see below), the current density j and plasma conductivity η are proportional to the electron density [107-108]:

$$j = \eta E = \frac{n_e e^2 E}{m_e N_0 \langle \sigma_t v \rangle}, \tag{15}$$

where E is the field ($E = V/l$, V is the voltage across the tube, and l is the length of the discharge gap), N_0 is the density of molecules in the ground state, $\langle \sigma_t v \rangle$ is the product, averaged over the electron velocity distribution, of the total effective cross section σ_t for the collision of electrons with molecules in the ground state times the electron velocity v. For molecular nitrogen σ_t is approximately equal to 10^{-15} cm^2, and the total cross section for inelastic processes is about 10^{-16} cm^2 [109, 110].

The relation between the current density and electron density can also be written in the form

$$j = en_e v_{\mathrm{dr}} \tag{16}$$

where v_{dr} is the drift velocity (mean directional velocity of electrons), which is a function of E/p (ratio of the field E to the gas pressure p) or, equivalently, of the function E/N_0. Equation (16) is better suited to calculations in view of the availability of experimental and theoretical data on the electron drift velocity in nitrogen [111, 112].

The temporal behavior of the current growth is described in general by the usual equation for an oscillatory loop:

$$L\frac{dJ}{dt} + R(t)J = V_0 - \int_0^t \frac{J}{C}\,dt, \tag{17}$$

in which

$$R(t) = \frac{l}{S\eta(t)} \tag{18}$$

is the resistance of the discharge gap due to the conductivity $\eta(t)$.

In the general case the resistance R(t) is determined by, in addition to spatial recombination processes, the processes immanent at the cathode. However, as shown in the preceding section, the role of cathodic processes can be neglected for a great many important cases, because with the formation of the cathode spot the current growth rate is limited only by spatial ionization processes, i.e., ionization by electrons and photoionization. At relatively low pressures such as are normally used in pulsed gas lasers the role of spatial photoionization processes should not be too great, and the current growth should therefore be governed for the most part by spatial ionization by electrons. The experimental data on the current growth rate corroborate this conclusion.

At not too high voltages, such that the current growth rate remains small and the inductive voltage drop is small, the current exhibits exponential growth with time (Fig. 28). This behavior on the part of the current is very simply explained on the basis of the assumption that the only source of spatial ionization is ionization by electrons.

In a real discharge the parameter E/p varies with time on account of the increase with time of the inductive voltage drop (with increasing dJ/dt) and the gradual discharging of the capacitance. However, in the case of the $N_2(1+)$ laser at not too high voltages and a sufficient capacitance neither of these factors begins to play a significant role until after termination of the laser emission pulse. Up to the instant of lasing cutoff the field in the tube can be regarded as time-invariant. The assumption of time invariance of E/p is tantamount to the stipulation

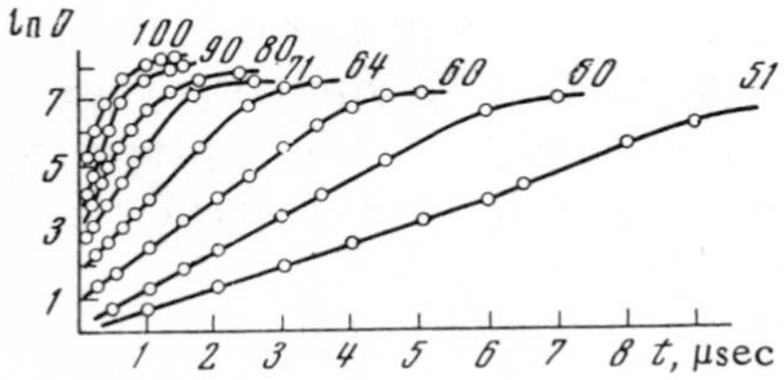

Fig. 28. Current growth versus time in the high-current phase of discharge. The numbers alongside the curves indicate the values of the parameter E/p (in V/cm · mm Hg). Tube bore, 2.5 cm; length, 3.4 m; p = 0.7 mm Hg.

that $C = \infty$ and $L = 0$. In this case we can write for the current density growth rate [differentiating (16)]

$$\frac{dj}{dt} = ev_{\text{dr}}\dot{n}_e, \tag{19}$$

and for $\dot{n}_e(t)$

$$\dot{n}_e = \alpha v_{\text{dr}} n_e, \tag{20}$$

$$n_e = n_e^0 e^{\gamma t}, \tag{21}$$

where

$$\gamma\left(\frac{E}{p}\right) = \alpha\left(\frac{E}{p}\right) v_{\text{dr}}\left(\frac{E}{p}\right) \tag{22}$$

is the product of the Townsend spatial ionization coefficient α, which is known from [113], and the drift velocity v_{dr} measured in [111]. From (21) and (16) we obtain

$$j(t) = j_0 e^{\gamma t}, \tag{23}$$

i.e., consistent with experiment, an exponential current growth. The dependence of the exponent γ on the parameter E/p as calculated according to Eq. (22) is in satisfactory agreement with the experimental data.

As shown by an analysis of the current and stimulated emission oscillograms, the assumption that $C = \infty$ and $L = 0$ can be adopted as a sensible idealization for the $N_2(1+)$ laser. This model is formally valid if prior to the instant of lasing cutoff $L(dJ/dt) \ll V_0$. For the $N_2(1+)$ laser the latter inequality holds over a fairly broad range of experimental conditions. The growth of the current and electron density in the discharge with time can be regarded as exponential, with time-independent exponential $\gamma = \alpha v_{dr}$. The initial electron density n_e^0 must be determined, wherever necessary, by the initial value of the total current.

Besides the electron density, it is also important to know the average electron energy ε_{av} (or effective temperature T_e) and velocity distribution of the electrons. In a weakly ionized plasma T_e is normally determined on the basis of the energy balance equations for 1 cm^3 of electron gas [10]:

$$\frac{d}{dt}\left(\frac{3}{2}kT_e n_e\right) = jE - \delta_{\text{eff}}\nu_{\text{eff}}\, n_e \cdot \frac{3}{2}k(T_e - T_g), \tag{24}$$

where T_g is the gas temperature (ordinarily $T_g \le T_e$) and ν_{eff} is the effective frequency of inelastic collisions in which the relative fractional energy lost is δ_{eff}. The product $\nu_{eff}\delta_{eff}$ is denoted in some papers as ν_u [109, 114].

For an exponential growth of electrons, T_e = const, and $\gamma \ll \nu_u$ the left term in (24) can be neglected. Ordinarily for $p = 0.75$ mm Hg the quantity $\nu_u = 5 \cdot 10^8$ sec^{-1} [109], and the left side of (24) becomes immaterial if $\gamma \le 5 \cdot 10^7$ sec^{-1}. The latter inequality holds when the parameter $E/p \le 200$ V/cm • mm Hg. In the experiments E/p did not exceed this figure as a rule.

With its left-hand side omitted, the balance equation (24) corresponds to the steady-state condition, for which there are theoretical [109] and experimental data [115] on the average electron energy. For $L = 0$ and $C = \infty$ the average electron energy can be considered to be approximately constant with time, provided that there are no significant changes in the state of the gas. Up to the instant of lasing cutoff this condition is met (see Chapter VII).

As for the electron velocity distribution $f(v)$, we know from [116] that the form of $f(v)$ is determined by elastic, inelastic, and interelectron collisions. As shown in [117], interactions between electrons reduce the deviation of $f(v)$ from a Maxwellian distribution. The role of these interactions in molecular nitrogen at p = 1 mm Hg becomes significant even for $n_e \sim 10^{12}$ cm^{-3} (the estimate is obtained from the ratio of the electron mean free paths with respect to inelastic and interelectron collisions [118]). A theoretical calculation of $f(v)$ for N_2 shows that even when electron–electron interactions are neglected, for $\varepsilon_{av} \sim 4$ eV the deviations from a Maxwellian distribution are relatively small [109]. This fact deserves attention. Angel [119] succeeded in calculating the Townsend ionization coefficient α in N_2 with reasonable accuracy on the basis of a Maxwellian distribution $f(v)$. However, this coefficient is extremely dependent on the form of $f(v)$. All this considered, it seems reasonably certain that the actual distribution $f(v)$ in molecular nitrogen in a high-current pulsed discharge must depart only slightly from Maxwellian, although, of course, the precise magnitude of the deviations is not known. The following consideration is important as far as we are concerned. Normally the deviations from a Maxwellian electron velocity distribution are greatest in the "tail" of the distribution. The excitation of triplet working states in the $N_2(1+)$ laser, on the other hand, is realized by electrons having energies of 7 to 10 eV [near the peak of $f(v)$], for which these deviations must not be too great. Consequently, the products $\langle \sigma v \rangle$, averaged over the Maxwellian distribution, must not differ too much from those which would be obtained on the basis of the true distribution.

It was mentioned above that fast electrons, accelerated by intense local potential gradients, can be generated in the discharge formation process. The role of such electrons in the filling of the laser levels is negligible. Thus, in the absence of strong local fields the fast electrons are "cooled" in a time of the order of the mean free transit time of this kind of electron in inelastic collision with a molecule. Under realistic conditions with $v_e \sim 10^8$ cm/sec, $\sigma_{in} = 10^{-16}$ cm^2 and $N_0 = 3.5 \cdot 10^{16}$ cm^{-3} this time turns out to be $\tau_{in} = 1/v_e \sigma_{in} N_0 \cong 3 \cdot 10^{-9}$ sec. In approximately the same time the average electron energy "emulates" the variation of the field in the discharge tube. These times are an order of magnitude smaller than the minimum experimentally observed duration of stimulated emission.

We now consider the contribution of diffusion and recombination processes to the current growth rate, as well as the influence of Coulomb collisions between electrons and ions on the conductivity of the pulsed-discharge plasma.

The lifetime of the neutral molecule due to free diffusion to the wall is equal to [114]

$$\tau = \left[D\left(\frac{2.4}{r_0}\right)^2 - \left(\frac{\pi}{e}\right)^2\right]^{-1}, \tag{25}$$

where

$$D = \frac{k\lambda}{3} = \frac{1}{3}\left(\frac{2kT}{M}\right)^{1/2}\frac{1}{N_0\sigma_y} \tag{26}$$

is the diffusion coefficient. For T = 300°K, $N_0 = 2.6 \cdot 10^{16}$ cm^{-3}, and $r_0 = 1$ cm the lifetime τ = 260 μsec, i.e., is clearly larger than the 0.2- to 4-μsec times used for the excitation of inversion.

Inasmuch as the Debye radius $r_D = 1.5 \cdot 10^3 (T_e/n_e)^{1/2}$ is already only $5 \cdot 10^{-2}$ cm for $n_e \sim 10^{10}$ cm^{-3} and $T_e \sim 10$ eV and as this value is much smaller than the tube radius (1 cm), the diffusion of charged particles to the tube wall can only be ambipolar. The ambipolar diffusion coefficient [114]

$$D_a = \frac{kT_e}{e} K^+ \tag{27}$$

is equal to $5 \cdot 10^3$ cm²/sec at a pressure of 0.75 mm Hg and $T_e = 3$ eV. The corresponding lifetimes of the charged particles due to ambipolar diffusion, τ_d, computed according to Eq. (25), are ~30 μsec. The electron losses due to ambipolar diffusion at the wall, n_e/τ_d, is two orders of magnitude smaller than the growth rate γn_e ($\gamma \lessapprox 10^6$ sec^{-1}) of the electron density, and these losses can be neglected.

Neither does the recombination of electrons play a major role in the current buildup. As we know [110], the recombination coefficient α_r is largest for relatively slow electrons ($T_e \sim$ 300°K). Under these conditions it is known to be equal to approximately $3 \cdot 10^{-7}$ cm³/sec for N_2. With an increase in the electron temperature α_r decreases inversely as $T^{3/2}$. Since the electron temperature in the N_2(1+) laser is about $3 \cdot 10^4$ °K, it is expected that α_r will have a maximum value of about 10^{-10} cm³/sec. Using this figure, we readily calculate that under typical conditions, with $n_e \cong 10^{13}$ cm^{-3} and $\gamma = 5 \cdot 10^6$ sec^{-1}, the growth rate of the electron density due to ionization is approximately three orders of magnitude greater than the rate of reduction of n_e due to recombination.

The only interest we have in Coulomb interactions between electrons and ions bears on the fact that these interactions govern the maximum degree of ionization of the plasma such that a linear relation still exists between the electron density and current density.

Taking Coulomb collisions of electrons with ions into account, we find for the current density [108]

$$j = \frac{n_e e^2}{m_e} E \frac{1}{1/\tau_{e0} + 1/\tau_{ei}}, \tag{28}$$

where $\tau_{e0} = 1/v_e N_0 \sigma$ is the time between collisions of an electron with neutral molecules and $\tau_{ei} = S \cdot 10^4 T_e^{3/2}/n_e$ is the time between Coulomb collisions of electrons with ions. It follows from (28) that the current density j is proportional to n_e under the condition $1/\tau_{e0} \gg 1/\tau_{ei}$. At a pressure of 1 mm Hg this condition fails only for $n_e \geq 3 \cdot 10^{14}$ cm^{-3}. The processes of immediate interest to us take place for much smaller electron densities, when the conductivity of the plasma is determined wholly by collisions between electrons and neutral molecules and j depends linearly on n_e.

CHAPTER VII

POPULATION DYNAMICS OF THE UPPER LASER LEVELS IN THE 1+ BAND SYSTEM OF N_2

§1. On the Excitation Mechanism for the Upper Laser Levels

When the time dependence of the absolute value of the electron density and effective electron temperature are known, it is a simple matter on the basis of the "direct electron impact" mechanism to calculate the time variation of the absolute value of the excitation rate of the upper laser state and the population of that state. Both of these variables can also be determined from the experimental data. A comparison of the measured and calculated data should reveal how closely the "direct electron impact" mechanism describes the excitation of inversion in the laser.

Given an exponential growth of the electron density and a constant electron temperature, and neglecting the radiative decay of the B state ($\tau_{\text{dec}} \cong 10$ μsec), we can write for the excitation rate of a particular vibrational level, say v' = 1, of the B state by electrons from the ground state

$$\dot{N}_{B_1} = N_0 n_e \langle \sigma_{X_0 B_1} v \rangle = N_0 \langle \sigma_{X_0 B_1} v \rangle \, n_e^0 e^{\gamma t}. \tag{29}$$

The population of this level is

$$N_{B_1} = N_0 \langle \sigma_{X_0 B_1} v \rangle \frac{n_e^0}{\gamma} e^{\gamma t}. \tag{30}$$

As apparent from (30), the population N_{B_1} must be, according to the mechanism of direct excitation of the B state, an exponential function of the time with exponent γ. The logarithm of the population, $\ln N_B$, will then be a linear function of the time, with slope γ, i.e., the experimentally measured time dependence of the logarithm of the B state population and the logarithm of the electron density must be parallel. This is indeed the situation observed in the experiments (Fig. 29), viz., the exponents for the current and spontaneous emission are roughly equal.

The excitation rate of the B_1 state, according to (29), must depend linearly on the electron density. This is in fact the dependence observed experimentally (Fig. 30).

The absolute values of the population of the B_1 state and excitation rate $\dot{N}_{B_1}$ calculated according to Eqs. (29) and (30) agree with the measured values to within ~1.3. (The absolute population of the B state was determined with the use of data on the absolute value of the population inversion at individual rotational transitions; see §2 of this chapter.)

Consequently, on the basis of the direct excitation mechanism for the laser levels not only is it possible to predict the exponential growth of the population of the important states with time, but also to calculate with acceptable accuracy the experimentally measured population rates of the laser levels and their absolute populations. This is the most significant quantitative evidence in favor of the "direct electron impact" mechanism in the N_2(1+) laser.

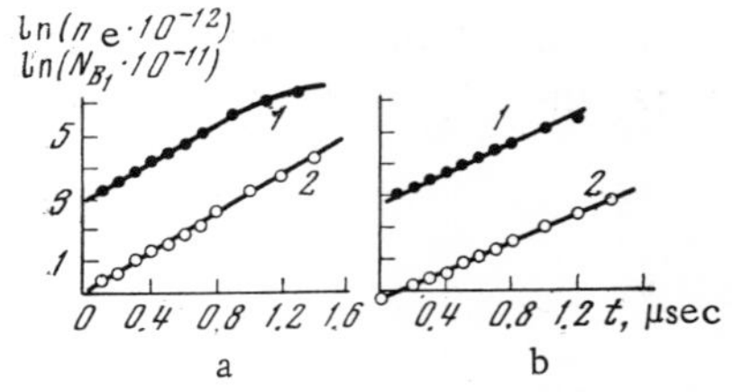

Fig. 29. Growth of the electron density (2) and B_1 state population (1) with time at voltages of 13.5 kV (a) and 12.2 kV (b). Tube bore, 1.5 cm; length, 2.0 cm; C = 0.12 μF; p = 0.8 mm Hg.

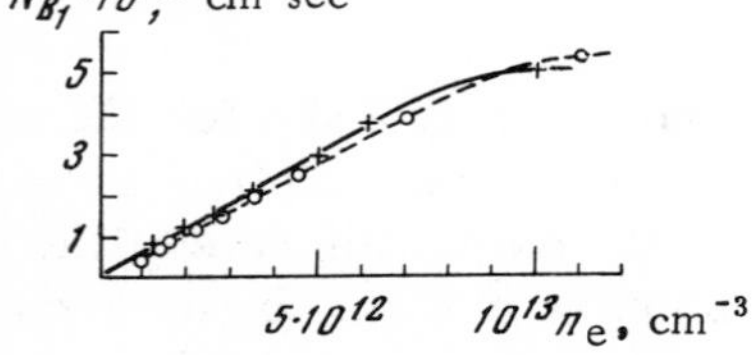

Fig. 30. Growth rate of B_1 state population versus electron density. Solid curve: 13.5 kV; dashed curve: 12.5 kV. Otherwise the nomenclature is the same as that of Fig. 29.

The direct excitation of the laser states is also witnessed by the following two considerations. The ratio $N_{v'}/f_{0v'}$, i.e., the measured relative population of the vibrational levels of the B state to the Franck–Condon factors $f_{0v'}$ for transition from the ground to the B state, which are known from [69], is approximately constant for different v' levels (0 < v' < 5). It is this dependence of the ratio $N_{v'}/f_0v_{0v'}$ in the direct electron excitation of the vibrational levels of the B state. The latter statement is readily understood with the realization that the quantity $\langle \sigma_{X B_{v'}} v \rangle$ entering into the expression for $N_{B_{v'}}$ [see (30)] is equal to $f_{0v'} \langle \sigma_{X_0 B} v \rangle$, where $\sigma_{X_0 B}$ is the total excitation cross section for the B state as a whole.

The second fact favoring the "direct electron impact" mechanism is the absence of any observable time shifts between the current and spontaneous emission pulses from the B state, even at relatively low gas pressures of ~0.1 or 0.2 mm Hg. This is the situation that must exist in the case of "direct" filling of the B state with electrons.

Thus, the now-known experimental facts confirm the "direct electron impact" mechanism.

It is essential to point out, however, that alternative channels are available for the filling of this state besides the "direct" excitation of the B state by electrons, in particular the mechanism of electron impact from some already excited state populated by direct impact from the ground state. It is easily demonstrated that in this mode of filling of the B state in the stage of exponential growth of the current N_B must be proportional to $e^{2\gamma t}$, as would be observed experimentally of course. Thus, $N_B \sim e^{\gamma t}$, as predicted by the "direct" population of the B state.

Another possible channel for filling of the B state is radiative decay from the C state. Inasmuch as the radiative lifetime of the C state is exceedingly small (~$5 \cdot 10^{-8}$ sec), it may be assumed that practically all the molecules excited by electrons in this state subsequently decay in the B state. The excitation rate of the B state by this channel is

$$(\dot{N}_B)_C = N_0 \langle \sigma_{0C} v \rangle \, n_e(t) \tag{31}$$

and the rate from the ground state is

$$\dot{N}_B = N_0 \langle \sigma_{0B} v \rangle \, n_e(t). \tag{32}$$

Under representative conditions $\langle \sigma_{0B} v \rangle$ is almost an order of magnitude greater than $\langle \sigma_{0C} v \rangle$, so that radiative decay from the C state yields only a small contribution to the total population of the B state (about 10%). The same conclusion is inferred from an analysis of the experimentally measured relative and absolute populations of the B and C states. Allowance for radiative decay from the C state is most important for the B_0 level, which is relatively weakly excited from the ground state. Radiative decay from the C state is responsible for up to 50% of the excitation of this level. For the B_1 level the contribution is already only 15%, and for B_2 it is only 10%.

§2. Critical Analysis of the Mechanism of Step-by-Step Filling of the Laser Levels

The hypothesis used in [3] and later in [19] of step-by-step excitation of the laser levels, when applied to the interpretation of lasing in the 1+ band system of N_2, contradicts both the experimental data presented here and the established information on the excitation cross section of the B state.

According to this hypothesis, in the very first phase of discharge mainly the singlet states of the nitrogen molecule and, less so, the triplet states are excited by electron impact. This assertion is based on the notion that the excitation of the triplet system proceeds by transition from the ground state with a change of spin and is therefore relatively improbable. The filling of the $B^3\Pi_g$ state, according to this hypothesis, obeys the following step-by-step process. First, direct excitation and cascades from higher singlet states account for the filling of the relatively long-lived levels of the $a^1\Pi_g$ state, whose radiative lifetime is 10^{-4} sec [83]. Then, collisions of the second kind between nitrogen molecules in an excited singlet state and molecules in the ground state cause an exchange of interaction energy, whereupon the triplet state $B^3\Pi_g$ state becomes populated.

The step-by-step filling hypothesis is based on the assumption that, on account of the change of spin, the effective cross section of the B state for excitation by electrons from the ground state must be small. This assumption is fallacious. The investigations [3, 19], in which lasing in the 1+ band system of N_2 is interpreted on the basis of the step-by-step excitation hypothesis, and, finally, [39], in which the hypothesis was first stated, were published before the cross section for excitation of the B state by electrons from the ground state had become known. This cross section was measured in [34, 35] and turned out to be very large. Its value (bearing in mind the maximum value) was compared with the total cross section for inelastic collisions of N_2 molecules with electrons, which is approximately 10^{-16} cm^2 [109]. This fact alone is a strong argument against the step-by-step filling hypothesis. It is augmented by a whole series of facts that further refute the hypothesis.

For a typical effective electron temperature of 3 to 8 eV the population rate of the singlet states of N_2, even for a comparable value of the effective cross section, $\sim 10^{-16}$ cm^2, must be considerably lower than for the triplet states. The truth of this statement is readily perceived with the realization that the maximum cross section for the excitation of triplet states is attained for a comparatively low energy ($\sim$10 to 15 eV), which approximately corresponds to the maximum of the electron energy distribution. In this near-threshold energy range the excitation cross sections of the singlet states are still very small. They become maximal only in the vicinity of $\sim$50 to 100 eV, i.e., in the far "tail" of the electron energy distribution, where the number of electrons is very much smaller than at the maximum. As a result, the population of the singlet states must be far weaker than that of the triplet states, and even if the transfer of excitation energy between the B state and singlet levels is possible, it cannot go from singlet to triplet states, but vice versa.

Also in support of the relatively weak excitation of singlet states in pulsed discharge is the fact that in the singlet band systems could not be detected in spontaneous emission over the entire range of investigated wavelengths from 2500 to 9000 Å.

Taking account of the resonance character of the transfer of excitation energy in collisions of heavy particles, the step-by-step filling hypothesis leads to the conclusion that the ratio N_v/f_{0v} must vary appreciably in transition from one level to another, having necessarily a very sharp maximum for $\Delta E = 0$. Thus, as noted above, the ratio $N_{Bv'}/f_{0v'}$ is virtually independent of v'.

In step-by-step excitation there must be between the current pulse and spontaneous emission from the B state a time shift approximately equal to the mean transit time of the molecule with respect to molecular collisions. This shift must increase as the pressure is lowered, amounting to about 0.5 μsec for $p = 0.2$ mm Hg. Such a shift was not observed experimentally. Besides the foregoing, the delay of the spontaneous emission pulse relative to the current must prevent the graph of $\dot{N}_{B_1} = f(n_e)$ from passing through zero. Again, this situation was not observed.

Consequently, the hypothesis of step-by-step filling of the B state is almost totally unsupported by any one experimental fact and contradicts the known data on the cross sections.

Within the framework of the "direct excitation" mechanism, on the other hand, the foregoing experimental results are readily explained, and the absolute population of the vibrational levels of the B state are calculated with acceptable accuracy on the basis of the cross sections measured in [34].

In light of the above discussion it seems logical to inquire into the explanation of the effect, frequently observed in several papers [39, 73, 120], for whose interpretation the step-by-step hypothesis was formulated in the first place. The main substance of this effect is the fact that when molecular nitrogen is irradiated by a narrow beam of electrons at a low gas pressure ($\sim 10^{-2}$ to 10^{-3} mm Hg) luminescence in the 1+ band system of N_2 is observed not only in the electron beam proper, but also far beyond the limits of the beam (out to ~ 5 mm from it [39]). This phenomenon is called the "spreading effect."

It is totally unnecessary to invoke such complex mechanisms as the mechanism of step-by-step excitation of the N_2 B state to account for the observed effect. The authors of [39] attempted on the basis of exactly this mechanism to interpret what had been observed. Actually, one need merely consider the comparatively large radiative lifetime of the B state in order to explain the spreading effect. This time can be calculated on the basis of information which, like the data on the excitation cross section for the B state, has been made available only recently with regard to the oscillator strength for the 1+ band system of N_2 and which was, of course, unknown to the authors of [39]. Judging from all they had to work with, they assumed that the radiative lifetime of the B state is very small. But if we consider the fact that the radiative lifetime of the B state is approximately 10 μsec, we readily show that under the conditions of [39] luminsescence in the 1+ band system should be observed to distances of 5 to 8 mm from the beam. Consequently, the spreading effect is satisfactorily explained just on the basis of the true radiative lifetime of the B state.

A second question that arises is, if we reject the hypothesis of "step-by-step" filling of the laser levels, how do we account for the stimulation of emission by the $N_2(1+)$ laser at the end of the excitation pulse? This is the aspect of the lasing process that was noted in [3]. It is interpreted as follows on the basis of the "direct electron impact" mechanism. Inasmuch as the threshold gain is not attained right away, but only after the buildup of inversion, lasing is always observed with a certain delay relative to the leading edge of the excitation pulse. For the $N_2(1+)$ laser these delays can be particularly great due to the smallness of the Einstein coefficient involved in the gain expression. This fact frequently leads to situations in which the excitation pulse width is comparable with the time interval required for the buildup of inversion. At the relatively low voltages used in [3] the buildup of the required inversion level takes about 1 μsec. The duration of the voltage pulse from the modulator was about the same. It is entirely understandable, therefore, that lasing should have been observed in [3] precisely at the end of the voltage pulse.

§3. Lasing Cutoff Mechanism; Simplified Laser-Level Diagram

In the preceding section we considered in detail the populating mechanism of the laser levels in the initial phase of the high-current stage of pulsed discharge, without discussing the processes leading to the cutoff of laser action with continued development of the discharge. Naturally, the most complete information about the lasing cutoff mechanism in the $N_2(1+)$ laser is afforded by an analysis of the experimental data. We carry out such an analysis in the present section in the example of the (1, 0) band.

The time variations of various experimentally measured variables essential to the understanding of the process culminating in inversion cutoff are shown in Fig. 31. The electron den-

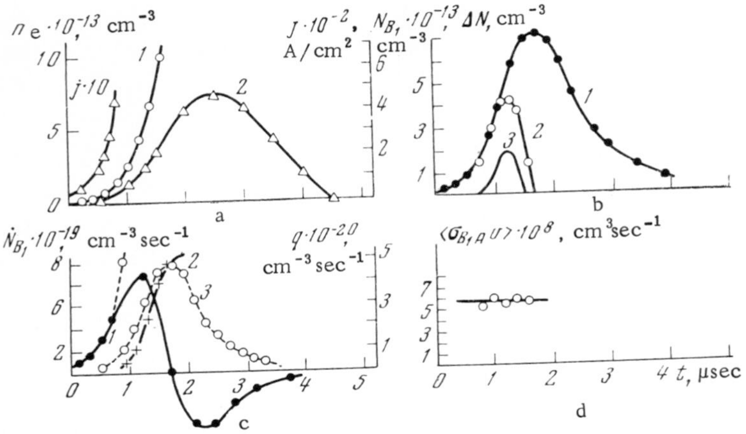

Fig. 31. Inversion excitation dynamics in the (1, 0) band of the 1+ band system of N_2. V = 13.5 kV; p = 0.8 mm Hg; C = 0.12 μF; tube bore, 1.5 cm; length of discharge gap, 2.3 m. a) Time variation of discharge current density (2) and electron density (1); b) time variation of absolute population of the B_1 state (1), inversion (2), and stimulated emission intensity (3); c) time dependence of rate of change of B_1 state population, $\dot{N}_{B_1}$ (1), filling rate of B_1 state from the ground state, $q = N_0 n_e(t) \langle \sigma_{X_0B_1} v \rangle_t$ (3), and de-excitation rate of B_1 state, $N_0 n_e \langle \sigma_{X_0B_1} v \rangle - \dot{N}_{B_1}$ (2) (right-hand ordinate); d) average of product $\langle \sigma_{B_1A} v \rangle$, determined according to experimental data.

sity n_e (curve 1 in Fig. 31a) was calculated from Eq. (16) according to the measured current value. The total inversion at the vibrational transition (1, 0) (curve 2 in Fig. 31b) was calculated on the basis of the inversion at the rotational $Q_1 9$ line and the condition $N_{v'} \gg N_{v''}$. The latter relation holds up to the time t_{max} at which the peak stimulated emission is attained. For $t > t_{max}$ the approximate variation of the total inversion at the transition is shown. The total inversion was calculated for a gas temperature of 320°K under the condition of complete rotational relaxation (see Chapter V). The rate of change dN_{B_1}/dt of the population of the B_1 state (curve 1 in Fig. 31c) was determined by graphical differentiation of the curve $N_{B_1} = f(t)$ (curve 2 in Fig. 31b). The excitation rate of the B_1 state from the ground level, $q = N_0 n_e(t) \langle \sigma_{X_0B_1} v \rangle$ (curve 3 in Fig. 31c) can be calculated from the measured dependence $n_e(t)$ under the condition $\langle \sigma_{X_0B_1} v \rangle = \text{const}$ over practically the entire time interval in which inversion exists (with the possible exception of the lasing cutoff time). The time dependence of the relative quantity $N_0 n_e(t) \cdot \langle \sigma_{X_0B_1} v \rangle_t$ was determined most accurately on the basis of data on the excitation rate of the C_2 state (the dependences of the cross sections $\sigma_{X_0B_1}$ and $\sigma_{X_0C_2}$ on the electron velocity are roughly identical [35]). For times $t \geq 1.5$ or 2 μsec, when n_e approaches 10^1 cm^{-3} and the role of step-by-step filling processes for the C state can no longer be neglected, curve 3 gives the approximate behavior of $q_1(t)$.

In the incipient phase of the high-current discharge stage, according to the direct electron impact mechanism, the quantity $\dot{N}_{B_1}$ increases exponentially with time (curve 1 in Fig. 31c). As shown in Chapter V, the lower laser levels in this case are relatively weakly populated. The absolute population of the B state in this phase coincides with the total inversion to within the experimental error.

With the further development of discharge the rate of change of the B_1-state population decreases to zero and then becomes negative, whereas the rate of excitation from the ground state attains a maximum (curves 1 and 3, respectively, in Fig. 31c). This result strongly indicates that, beginning with a certain time, a process of intense de-excitation of the B_1 state sets in and, simultaneously, judging from the reduction of the inversion (curve 2 in Fig. 31b), rapid filling of the lower laser state A_0 is initiated. The excitation dynamics of the laser levels in this phase of discharge is satisfactorily explained by taking account of the de-excitation of the upper laser state by electrons in "superelastic collisions" with subsequent transition of the molecule into the A state:

$$N_2(B_1) + e \rightarrow N_2(A) + e + \Delta E. \tag{33}$$

Estimates show that for typical values T_e = 3 to 8 eV the de-excitation rate of the B state by electrons to the ground state is two orders of magnitude smaller and by ionization is approximately one order of magnitude smaller than for transitions to the A state. Due to the small radiative lifetime of the C state practically all the molecules excited by electrons from the B state to the C state revert to the former. If we assume that de-excitation of the B_1 level takes place only in the process (33) (the contribution of radiative decay can be neglected), we can write for the rate $\dot{N}_{B_1}$

$$\dot{N}_{B_1} = N_0 \langle \sigma_{X_0 B_1} v \rangle n_e - N_{B_1} \langle \sigma_{B_1 A} v \rangle n_e, \tag{34}$$

whence

$$\langle \sigma_{B_1 A} v \rangle = \frac{N_0 \langle \sigma_{X_0 B_1} v \rangle n_e - \dot{N}_{B_1}}{N_{B_1} n_e}, \tag{35}$$

where $\sigma_{B_1 A}$ is the total effective cross section for de-excitation of the B state by electrons to all lower-lying vibrational levels of the A state. The values of $\langle \sigma_{B_1 A} v \rangle$ calculated according to Eq. (35) are given in Fig. 31d.

The quantity $\langle \sigma_{B_1 A} v \rangle$ is easily calculated by means of the Bethe equation [60] using the oscillator strength from [86] for the 1+ band system of N_2. The value thus obtained, $2 \cdot 10^{-8}$ cm^3/sec, coincides, correct to a factor of ~3, with the value calculated from Eq. (35). The value of $\langle \sigma_{B_1 A} v \rangle$ calculated according to (35) is approximately constant for different times (Fig. 31d), as it should be in the de-excitation of the B state in process (33). Consequently, the experimental results attest to the fact that the main process of de-excitation of the upper working levels leading to the cutoff of inversion in the N_2(1+) laser are "superelastic" collisions of electrons with excited molecules, Eq. (33).

It must be particularly emphasized that our experimental conditions were such that up until the time of inversion cutoff we were able to neglect the influence of discharging of the capacitor and the voltage drop across the inductance or other impedance in the working circuit. Under different conditions these factors could exert a profound influence on the excitation dynamics of the laser states, compounding the difficulty of analyzing the physical processes involved.

In the analysis of the excitation and de-excitation of the upper and lower laser vibrational states in process (33) it is required, in general, to take account of the transitions to all possible vibrational levels. As shown by a more detailed analysis, however, a relatively simple excita-

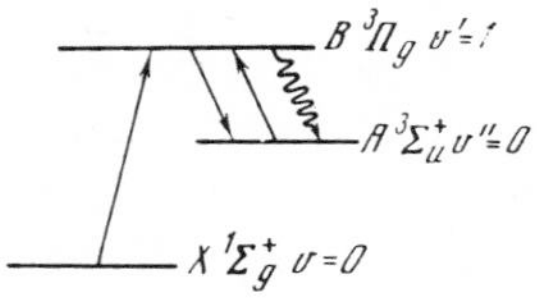

Fig. 32. Simplified inversion excitation diagram in the (1, 0) band of the N_2(1+) laser.

tion diagram is applicable (Fig. 32). As a matter of fact, for the strongest laser emission bands (1, 0) and (2, 0), as well as for all other bands to a certain extent, the total de-excitation rate of the upper laser level by electrons, $\sum_{v''} N_{v'} n_e \langle \sigma_{v'v''} v \rangle$, is approximately equal to the total population rate v'' of the lower laser level as determined by the expression $\sum_{v'} N_{v'} n_e \langle \sigma_{v'v''} v \rangle$. The latter fact makes possible the given simplification. For superelastic collisions, of course, the total cross section is used in this case. This procedure indirectly accounts for the action of all the vibrational levels.

As we saw in Chapter III, molecular heating and dissociation play a major role in the excitation dynamics of the molecular states of the H_2 laser. For the N_2(1+) these processes, all things considered, do not afford a sizable contribution. Of course, heating and dissociation of the molecular gas also take place in the N_2 discharge, but at the time of inversion cutoff they have not yet developed to the required extent.

The comparatively slow development of the foregoing processes in N_2, by contrast with H_2, is explained by a great many factors. The heating of the neutral gas by electrons in the nitrogen case is roughly an order of magnitude weaker than in hydrogen, because the ratio m/M for N_2 is considerably smaller than for H_2. Another heating channel, i.e., via collisions with ions, is also about one order of magnitude weaker for N_2 on account of the lower ion drift velocity. Finally, a mode of heating and dissociation important in the H_2 case, namely dissociation by direct electron impact through the excitation of unstable molecular states, is generally absent in N_2. According to data given in, for example, [28], no significant dissociation of N_2 molecules in electronic excitation is observed up to the ionization potential. A special calculation of the heating of N_2 in pulsed discharge on the basis of the exponential growth of the electron density and with regard for the two heating factors mentioned above shows that the gas is heated by about 50°K at the time of lasing cutoff. This heating is very slight. As we saw in Chapter V, it also follows from the laser emission intensity distribution among the lines of the Q_3 branch that the gas is close to room temperature at the instant of lasing. In this connection nothing can be said about thermal dissociation in the N_2 laser, at least up until the time of lasing cutoff. Dissociation by electrons, as we saw above, must also be minor. Under these conditions the only serious cause of population reduction of the ground state comprises inelastic collisions between electrons and molecules. If we proceed from the total inelastic impact cross section of 10^{-16} cm^2 and calculate the fraction of ground-state molecules experiencing inelastic collisions with electrons, we find that at the time of lasing cutoff the number of molecules in the ground state can be diminished by about 10%. In accordance with the foregoing discussion the assumption of a constant concentration of molecules in the ground state N_0 = const seems to be a fully realistic idealization for the calculations.

The authors of [121] have also inferred the necessity of taking account of "superelastic" collisions of electrons with excited molecules in the upper working state of the N_2(2+) laser. The results of this paper indirectly support the validity of our conclusion regarding the inversion cutoff mechanism in the N_2(1+) laser.

CHAPTER VIII

ELEMENTARY QUANTITATIVE THEORY OF LASER ACTION IN THE 1+ BAND SYSTEM OF N_2

§1. Population Inversion Dynamics

For the simplified level diagram shown in Fig. 32 we can at once write down the population equations for the laser states in a given emission band in the amplification mode, for example the (1, 0) band. In accordance with the discussion of the preceding chapter the growth of the electron density with time is considered to be exponential, while the electron temperature and, together with it, all products σv averaged over a Maxwellian electron velocity distribution, as well as the quantity γ are assumed to be independent of the time. The radiative decay of the upper laser level is neglected. In this case the population equations for the vibrational states B_1 and A_0, i.e., the upper and lower laser levels of the (1, 0) band (abbreviated B and A, respectively) have the following form in the amplification mode:

$$\dot{N}_B = N_0 n_e^0 \langle \sigma_{XB} v \rangle e^{\gamma t} + N_A n_e^0 \langle \sigma_{AB} v \rangle e^{\gamma t} - N_B n_e^0 \langle \sigma_{BA} v \rangle e^{\gamma t}, \qquad (36)$$

$$\dot{N}_A = N_B n_e^0 \langle \sigma_{BA} v \rangle e^{\gamma t} - N_A n_e^0 \langle \sigma_{AB} v \rangle e^{\gamma t}, \qquad (37)$$

where σ_{XB}, σ_{AB}, and σ_{BA} are the effective cross sections for electronic excitation of the B_1 level from the ground state, of the B_1 level from the A_0 level, and of the A_0 level from the B_1 level, respectively; and N_B and N_A are the populations of the upper (B_1) and lower (A_0) laser states. For brevity we denote $N_0 n_e^0 \langle \sigma_{XB} v \rangle = a$, $n_e^0 \langle \sigma_{AB} v \rangle = b$, $n_e^0 \langle \sigma_{BA} v \rangle = c$. In this notation Eqs. (36) and (37) assume the form

$$\dot{N}_B = a e^{\gamma t} + N_A b e^{\gamma t} - N_B c e^{\gamma t}, \qquad (38)$$

$$\dot{N}_A = N_B c e^{\gamma t} - N_A b e^{\gamma t}. \qquad (39)$$

After some straightforward transformations we obtain the following simple analytic expressions for the population of the lower and upper laser states:

$$N_A = \frac{a}{\gamma}(e^{\gamma t} - 1) - N_B, \qquad (40)$$

$$N_B = \frac{ac}{\gamma (b+c)}(e^{\gamma t} - 1) - \frac{ac}{(b+c)^2}\{1 - e^{-\frac{b+c}{\gamma}(e^{\gamma t}-1)}\}, \qquad (41)$$

and for the inversion $\Delta N = N_B - (g_B/g_A) N_A'$:

$$\Delta N = \left(1 + \frac{g_B}{g_A}\right)\frac{ac}{(c+b)^2}\{1 - e^{-\frac{b+c}{\gamma}(e^{\gamma t}-1)}\} - \frac{a}{\gamma (b+c)}\left(c\frac{g_B}{g_A} - b\right)(e^{\gamma t} - 1). \qquad (42)$$

From the maximum condition for ΔN we readily find the time t_{max} at which the inversion reaches a maximum and the absolute value of the total inversion $(\Delta N)_{max}$ between the laser levels:

$$t_{max} = \frac{1}{\gamma}\ln\left[\frac{\gamma}{b+c}\ln\left\{\frac{c(1+g_B/g_A)}{c(g_B/g_A) - b}\right\}\right], \qquad (43)$$

$$(\Delta N)_{max} = \frac{N_0 \langle \sigma_{XB} v \rangle}{\langle \sigma_{AB} v \rangle} k_1\left(\frac{\varepsilon_{BA}}{kT_e}\right). \qquad (44)$$

In expression (44) the factor $k_1(\varepsilon_{BA}/kT_e)$ is a moderately varying function of the parameter E/p:

$$k_1\left(\frac{\varepsilon_{BA}}{kT_e}\right) = k(\delta) = \frac{\left\{1 - \frac{(e^\delta - 1)}{1 + (g_A/g_B)\, e^\delta} \ln\left[\frac{e^\delta (g_A/g_B + 1)}{e^\delta - 1}\right]\right\}}{1 + (g_A/g_B)\, e^\delta}. \quad (45)$$

In the derivation of Eqs. (43) and (44) we used the relationship between the average direct and reverse cross sections:

$$\langle \sigma_{AB} v \rangle = \frac{g_B}{g_A} e^{-\varepsilon_{BA}/kT_e} \langle \sigma_{BA} v \rangle.$$

It is very instructive to compare the results of calculations according to Eqs. (42)-(44) with the experimental data. This kind of comparison is important for the experimental verification of the initial assumptions, as well as the physical models used in the derivation of Eqs. (42)-(44).

For the $N_2(1+)$ laser the time t_{max} of maximum inversion must be approximately equal to the time of maximum stimulated emission (this conclusion has been confirmed experimentally for small values of E/p). The latter time can be measured easily and accurately, and then compared with the calculations according to Eq. (43). The calculated dependence of t_{max} on the parameter E/p is shown in Fig. 10 in Chapter IV (dashed curve 1), along with the results of a measurement of the peak-emission time (solid curve 1). In the calculation of t_{max} the quantity $\langle \sigma_{AB} v \rangle$ was determined on the basis of the experimental value $\langle \sigma_{BA} v \rangle = 6 \cdot 10^{-8}$ cm^3/sec [see Fig. 31d and Eq. (35)], and the initial electron density $n_e^0 = 10^{-11}$ cm^{-3} was calculated from the initial value of the current. The dependence of $\langle \sigma_{AB} v \rangle$ on the average electron velocity was determined from the Bethe approximation using the effective Gaunt factor [122] and a Maxwellian electron velocity distribution. As evident from Fig. 10, satisfactory agreement with experiment is observed not only for the absolute value of t_{max} for a fixed value of E/p, but also for the entire dependence $t_{max} = f(E/p)$. For small values of E/p the measured and calculated values coindide to within 10 to 20%, and for larger values of the parameter to within 50%. Moreover, Eq. (43) makes it possible to predict the variation of the relative laser pulse width Δt_L as a function of E/p (curve 2 in Fig. 10). The calculated values of Δt_L, normalized to the experimental data at E/p ~ 70, are given by the dashed curve.

The agreement of the calculated and measured values affords conclusive evidence in support of the assumptions used in the derivation of Eq. (43). It must be pointed out that, since the inductive voltage drop was neglected throughout the calculations, agreement with the experimental was expected for not too large values of the parameter E/p. It is possible that the good agreement with the experimental over the entire investigated range of variation of E/p is attributable to mutual cancellation of diverse effects elicited by the inductive voltage drop for large values of E/p (we are thinking primarily in terms of the reduction in growth rate of the electron density and the decrease in the effective electron temperature with time due to the inductive voltage drop).

The calculation of the total inversion $\Delta N_{v'v''}$ at the vibrational transition (v', v'') and of the maximum inversion $(\Delta N_{v'v''})_{max}$ according to Eqs. (42) and (44) is elementary. However, the experimental measurement of these variables meets with major difficulties. To facilitate the comparison with experiment it is expedient to calculate the gain for the individual rotational line. Proceeding from (40) and (41), we readily determine the total populations of the upper and lower vibrational levels as a function of the time. Then, taking account of the fact that the distribution of molecules in the rotational states is thermal, with a gas temperature T_g, we easily obtain an expression for the inversion $\Delta N_{J'J''} = N_{J''} - N_{J''} g_e'[2J'+1)/(2J''+1)]$ at the rotational transition $Bv'J'\Lambda' \rightarrow Av''J''\Lambda''$, abbreviated J' → J'' ($g_e'$ is the statistical weight of the upper electron state). We need merely invoke Eqs. (40) and (41):

$$\Delta N_{J'J''} = g_N(2J'+1)\frac{\exp(-\varepsilon_{J'}/kT_g}{Z_{v'}}\left\{\left(1+\frac{g_B}{g_A}\chi\right)\frac{ac}{(c+b)^2}\times\right.$$
$$\left.\times\left(1-e^{-\frac{b+c}{\gamma}(e^{\gamma t}-1)}\right)-\frac{a}{\gamma(b+c)}\left(c\frac{g_B}{g_A}\chi-b\right)(e^{\gamma t}-1)\right\}, \quad (46)$$

in which $\chi = \chi(T_g, J', J'')$ is given by the expression

$$\chi(T_g, J', J'') = \frac{Z_{v'}}{Z_{v''}}\exp\left(-\frac{\varepsilon_{J''}-\varepsilon_{J'}}{kT_g}\right). \quad (47)$$

Now it is a simple matter to find an equation for the temporal-maximum inversion at the individual rotational lines:

$$(\Delta N_{J'J''})_{\max} = \frac{N_0\langle\sigma_{XB}v\rangle}{\langle\sigma_{AB}v\rangle} g_N(2J'+1)\frac{e^{-\varepsilon J'/kT_g}}{Z_{v'}} k_2\left\{\frac{\varepsilon_{BA}}{kT_e}, \chi\right\}, \quad (48)$$

in which $_{\max} k_2\{\varepsilon_{BA}/kT_e\ \chi\} = k_2(\delta, \chi)$ is a slowly varying function of E/p:

$$k_2(\delta,\chi) = \frac{1-\dfrac{(\chi e^{\delta}-1)}{1+(g_B/g_A)e^{\delta}}\ln\left\{\dfrac{e^{\delta}[1+(g_A/g_B)\chi]}{\chi e^{\delta}-1}\right\}}{1+(g_A/g_B)e^{\delta}}. \quad (49)$$

Correct to a factor of ~1.5 to 2, the maximum absolute value of the inversion at the rotational transition is equal to the total inversion at the vibrational transition, multiplied by the relative population of the upper rotational level.

From Eqs. (9) and (48) we readily obtain an expression for the gain at the individual rotational line:

$$(k_{J'J''})_{\max} = \frac{16(\pi^5\ln 2)^{1/2}}{3hc}\frac{\nu}{\Delta\nu_g}\frac{S_e^{BA}}{g_e'}q_{v'v''}S_{J'J''}\frac{N_0\langle\sigma_{XB}v\rangle}{\langle\sigma_{AB}v\rangle}g_N(2J'+1)\frac{e^{-\varepsilon_{J'}/kT_g}}{Z_{v'}}k_2\left\{\frac{\varepsilon_{BA}}{kT_e}, \chi(T_g, J', J'')\right\}. \quad (50)$$

The gain calculated according to Eq. (50) at the rotational line $Q_1 9$ is compared in Fig. 33 with the experimental value. If we proceed from the experimentally determined value of $\langle\sigma_{BA}v\rangle$, the calculated gain coincides with the measured value to within 30 to 50%. On the basis of Eq. (50) we can also predict the approximate variation of $(k_{J'J''})_{\max}$ as a function of the parameter E/p. The expression for the gain as a function of the time is analogous to Eq. (42). The results of calculations of this dependence, $k_{J'J''} = f(t)$, are compared in Fig. 34 with the experimental data. We see that the general behavior of the curves with time roughly coincides.

Thus, not only $t_{\max}$, but even the absolute value of the gain $k_{J'J''}$ at the individual rotational transition can be calculated with acceptable accuracy. In addition, the time dependence of $k_{J'J''}$ and the dependence of $(k_{J'J''})_{\max}$ on E/p can also be calculated with quantitative accuracy. In all of this we find further proof of the validity of the assumptions underlying the calculations, specifically the mechanism of inversion by "direct electron impact" and the adopted physical model of pulsed discharge.

From (44) we readily obtain an expression for the maximum attainable total energy accumulated in the vibrational transition $(W_{v'v''})_{\max}$ [Eq. (44) is multipled by $h\nu$ and the factor $g_A/(g_A+g_B)$]:

$$(W_{v'v''})_{\max} = \frac{N_0\langle\sigma_{XB}v\rangle}{\langle\sigma_{AB}v\rangle}h\nu k_1\left(\frac{\varepsilon_{BA}}{kT_e}\right)\frac{g_A}{g_A+g_B}. \quad (51)$$

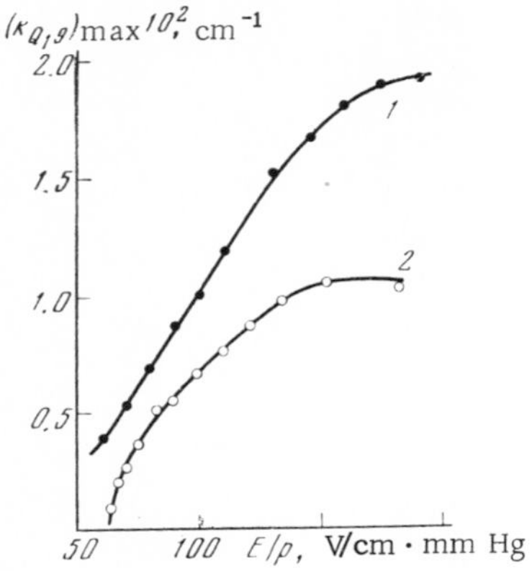

Fig. 33. Calculated (1) and measured (2) gain at the $Q_1 9$ line of the (1, 0) band versus parameter E/p. The measurements were carried out in a tube of length 2.0 m and bore 1.5 cm at an N_2 pressure of 0.75 mm Hg.

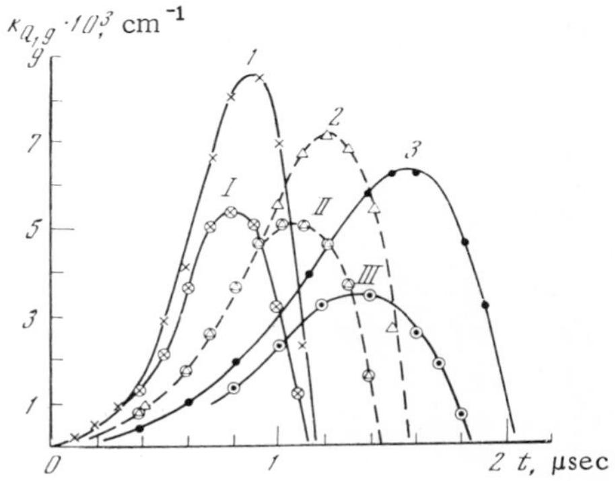

Fig. 34. Gain at the $Q_1 9$ line versus time for various discharge voltages. 1-3) Calculated; I-III) measured at voltages of 14.5, 13.5, and 12.2 kV, respectively. The experimental conditions are the same as in Fig. 33.

As apparent from (51), the energy W_{max} is determined principally by three factors: the gas pressure, the excitation cross section for the upper laser level, and the "superelastic" collision cross section. The expression for W_{max} contains neither the quantity γ nor the initial electron density n_e^0. At a pressure of 1 mm Hg the maximum total inversion at the three vibrational transitions v' = 0, 1, and 2, calculated according to Eq. (51), is approximately (5 or 6) $\cdot 10^{14}$ cm^{-3}, which corresponds to an energy stored per 1 cm^3 of about $5 \cdot 10^{-5}$ J.

In concluding this section, we point out that analogous results can be obtained if the reverse excitation of the B state by electrons from the C state is disregarded in the laser-level population diagram (see Fig. 32). The inclusion of this process merely contributes to a more precise description of the inversion at cutoff. Consequently, the reduction in the present article of the whole set of vibrational levels to two is essentially one of the possible techniques for a more accurate calculation of the inversion at that time.

We now look further into the problem of criteria for judging the applicability of the given physical model of the plasma. In calculations of the parameters of the active medium the role of the inductance L, internal resistance r, and discharging of the capacitance C can be neglected if the following conditions hold at the time of stimulated emission: $L\frac{dJ(t)}{dt} \ll V_0$, $rJ(t) \ll V_0$, $\int_0^t \frac{J(t)\,dt}{C} \ll V_0$. In this case it may be assumed that L = r = 0 and C = ∞. For an exponential growth of the current the quantities J(t) and dJ(t)/dt at the time t_{max} of peak inversion are equal to $J(t_{max}) \cong 0.7\gamma/\langle\sigma_{B'A}v\rangle$, and $dJ(t_{max})/dt \cong 0.7\gamma^2/\langle\sigma_{B'A}v\rangle$. The main experimental results were obtained for a relatively slow buildup of discharge (E/p ~ 70 V/cm • mm Hg), when the inequalities written above have the following form at t_{max}: $L \ll 20\ \mu H$, $p \ll 45\ \Omega$, and $C \gg 0.02\ \mu F$. We recall that the corresponding parameters of the discharge loop are L = 2 or 3 μH, r ~ 1 Ω, and C = 0.12 μF. The role of the inductance is not fundamentally significant for the $N_2(1+)$ laser. Even at a high current growth rate (E/p = 200 to 300 V/cm • mm Hg) the attainable circuit inductance is about 100 nH. We note that excitation systems are now available with an inductance two orders of magnitude smaller than this value (~1 nH) [25].

§ 2. On the Ultimate Laser Power

It is extremely relevant to compute the power density of the laser at saturation, relying on the same assumptions as in the gain calculations.

Following [44] and making use of the abbreviated notation introduced in the preceding section, we obtain the following equations for the population of the laser levels for the (1, 0) band at saturation:

$$\dot{N}_B = ae^{\gamma t} + N_A be^{\gamma t} - N_B(A + ce^{\gamma t}) - R_{BA}, \tag{52}$$

$$\dot{N}_A = N_B(ce^{\gamma t} + A) - N_A be^{\gamma t} + R_{BA}, \tag{53}$$

$$N = N_B \sim N_A \frac{g_B}{g_A}, \tag{54}$$

in which $R_{BA} = R_{BA}(N_B - (g_B/g_A)N_A)$ is the stimulated emission photon density. In Eqs. (52) and (53) we have introduced for greater generality the radiative decay from the C state with probability A. For the laser power yielded at saturation from 1 cm³ of the laser volume we obtain

$$P_s = h\nu\left\{\frac{ag_A}{g_A+g_B}\left(1 - \frac{g_A}{g_B}\frac{A}{\gamma}\right)(e^{\gamma t}-1) - \frac{g_A}{g_B+g_A}\frac{ab}{\gamma}\left(e^{\varepsilon_{BA}/kT_e}-1\right)\left(e^{2\gamma t}-1\right)\right\}. \tag{55}$$

The maximum power P_{max} is

$$P_{max} = \frac{\gamma N_0\langle\sigma_{XB}v\rangle}{\langle\sigma_{AB}v\rangle} h\nu\xi\left(\frac{\varepsilon_{BA}}{kT_e}\right), \tag{56}$$

where $\xi(\varepsilon_{BA}/kT_e)$ is given by the expression

$$\xi\left(\frac{\varepsilon_{BA}}{kT_e}\right) = \frac{g_A}{g_A+g_B}\frac{[1-(g_B/g_A)(A/\gamma)]^2}{4\left(e^{\varepsilon_{BA}/kT_e}-1\right)}. \tag{57}$$

As evident from Eq. (56), the maximum specific stimulated emission power of the $N_2(1+)$ laser, as opposed to the population inversion, is proportional to the factor γ, which increases rapidly with the parameter E/p. For p = 1 mm Hg and E/p = 100 V/cm • mm Hg we find that P_{max} is about 0.5 kW/cm³ for the (0,1) band and about 1.5 kW/cm³ for the three strongest bands, (0, 0), (1, 0), and (2, 1). Under the stated conditions this figure represents the maximum attainable laser power without Q switching.

At first glance, the experimental results seem to contradict the foregoing conclusions. The value of the peak power obtained experimentally is much smaller than the values cited. The experimentally observed behavior of the voltage dependence of the laser power does not even remotely resemble what is expected on the basis of Eq. (56). Whereas according to the latter equation the laser power should increase abruptly with the voltage (primarily due to the factor γ), in the experiments we observe an initial growth followed by saturation, and then even an inflection of the laser power curve as the voltage is increased.

These discrepancies are readily grasped, however, if we take account of the finite buildup time of the photon avalanche process in the $N_2(1+)$ laser. Thus, the equation for the growth of the photon density n_{ph} per mode has the form

$$\dot{n}_{ph} = n^0_{ph} + (\alpha - k)\, c n_{ph}, \tag{58}$$

in which n^0_{ph} is the photon density per mode due to radiative decay, $\alpha - k$ is the effective gain with regard for losses, and c is the velocity of light. For n^0_{ph} we can write

$$n^0_{ph} = \frac{N_{B_J} A_J}{\Delta\nu_D} \frac{1}{8\pi\nu^2/c^3}, \tag{59}$$

where the left factor is the number of photons emitted at one rotational line per unit frequency interval $\Delta\nu$ and the denominator of the right factor is the number of modes per cubic centimeter and unit frequency interval. Integrating Eq. (58), we obtain

$$n_{ph}(t) = \frac{n^0_{ph}}{(\alpha - k)\, c} \left(e^{(\alpha-k)ct} - 1\right). \tag{60}$$

We have assumed here that the gain and, hence, the inversion at the laser transition are independent of the time.

If a value of $\sim 10^{16}$ cm^{-3} s adopted for the photon density at saturation, we find for the strongest rotational lines that the time t required for buildup of the photon avalanche from the spontaneous noise level to the saturation power ($\alpha - k = 10^{-2}$ cm^{-1}) is $t = 10^{-7}$ sec. Consequently, even for the strongest rotational lines, whose gain is the maximum, the buildup time of the photon avalanche is comparable with the inversion holding time for $E/p = 100$ V/cm • mm Hg. As for the remaining bulk of the rotational lines that have a considerably smaller gain but still contribute heavily to the main power, they cannot develop a significant photon avalanche, and inversion is not realized for them.

Coupled with the limited value of the gain with increasing voltage, it is the finite buildup time of the photon avalanche that accounts for the experimentally observed saturation and even the inflection of the peak power curve as the parameter E/p is increased.

The potential capabilities of a laser acting at the infrared transitions of N_2 are comparable with those of the most powerful pulsed gas-discharge lasers acting at the ultraviolet transitions of the same molecule. However, due to the considerably lower transition probability and, as a result, the slower buildup of the photon avalanche process these capabilities are not realized for the infrared transitions under normal conditions. It is reasonable to suppose that the possibilities of the $N_2(1+)$ laser will be more fully appropriated by techniques for the amplification of short light pulses from a special driving generator. In this event one of the most important assets of the $N_2(1+)$ laser would be brought fully into its own, namely the possibility of the accumulation of a relatively large supply of active molecules and their subsequent rapid de-excitation. In this case we can hope for a sizable gain in the peak power of the laser and the utilization of a major portion of the cumulative inversion.

Besides the laser power, it is also important to calculate the energy of the laser emission pulse at saturation. For this we need merely integrate expression (55) up to the time t_k at which $P_s = 0$. As shown by numerical analysis, the latter condition occurs when $e^{\gamma t} \gg 1$.

For this case the expression for t_k has the form

$$t_k = \frac{1}{\gamma} \ln \left\{ \frac{\gamma}{b} \frac{1 - (g_B/g_A)(A/\gamma)}{e^{\varepsilon_{BA}/kT_e} - 1} \right\}.$$

The total emission pulse energy W_s at saturation per 1 cm^3 in the (1, 0) band is given by the expression

$$W_s = \frac{N_0 \langle \sigma_{XB} v \rangle}{\langle \sigma_{AB} v \rangle} h\nu \frac{g_A}{g_B + g_A} \frac{[1 + (g_B/g_A)(A/\gamma)]^2}{2 (e^{\varepsilon_{BA}/kT_e} - 1)}. \quad (62)$$

As we see, this expression coincides (up to a weakly varying function of E/p, or a factor of ~4 to 6) with the expression (51) for the energy stored in a vibrational transition in the amplification mode. As in amplification, the laser emission energy density in the saturation mode does not depend either on the current growth rate or on the initial electron density. At a pressure of 1 mm Hg the total energy W_s of the laser pulse at saturation in the (0, 0), (1, 0), and (2, 1) bands is about four to six times its value in amplification, amounting to approximately (1 or 2) $\cdot 10^{-4}$ J/cm^3.

The foregoing analysis of the operation of the N_2(1+) laser in the amplification and saturation modes is essentially based on only the most general considerations. The results of this analysis can be applied to any laser of this particular category, whether the laser uses an atomic gas, ions, or metal vapor, provided only that the conditions under which the calculation is carried out are indeed fulfilled for the given laser. On the basis of the expressions deduced above for the laser pulse energy we can, in particular, attempt to suggest methods for enhancing the stimulated-emission energy in pulsed gas lasers similar to the N_2(1+) and to learn more about the ultimate energy of the emission pulse for lasers of this type.

Equations (51) and (62) for the stimulated-emission energy W may be written as follows:

$$W = \frac{N_0 \langle \sigma_{01} v \rangle}{\langle \sigma_{21} v \rangle} h\nu k', \quad (63)$$

where the indices 0, 1, and 2 enumerate, respectively, the ground and upper and lower laser states, and k' is a certain factor that varies slightly with the parameter E/p. For the 1+ band system of N_2 this factor is equal to 0.15 in the amplification mode and to 0.6 to 1.0 in the saturation mode. Equation (63) should be valid for a relatively large radiative lifetime of the upper levels and meager population of the lower state with electrons from the ground levels. An intense radiative decay of the upper laser level, as is apparent, for example, from (62), merely reduces the energy W.

As we see from (63), there are four principal factors involved in raising the specific energy W of stimulated emission:

1. It is required to choose lasers having a large excitation cross section for the upper working level. A cross section 10^{-16} cm^2, with $\langle \sigma_{01} v \rangle$ of order 10^{-8} cm^3/sec, is not all out of reason. For the N_2(1+) laser the quantity $\langle \sigma_{01} v \rangle$ for the upper working state, as calculated from the known value of the cross section σ_{SB} according to [34, 35], closely approaches the stated value, but not all the vibrational levels are "operative," so that the "active" value of $\langle \sigma_{01} v \rangle$ is approximately 10^{-9} cm^3/sec.

2. The energy of the stimulated emission pulse must increase for systems having a smaller value of $\langle \sigma_{21}, v \rangle$. In the N_2(1+) laser this quantity, which attains 10^{-7} cm^3/sec, is more than two orders of magnitude greater than the cross section for "superelastic" collisions

...d state. The maximum value of the ratio $\langle\sigma_{01}v\rangle/\langle\sigma_{21}v\rangle$ is determined ...pending on the ratio of the statistical weights, can range from 0.1 to

...gy quantum $h\nu$ can be comparable with the average electron energy (5 to 10

4. The specific stimulated emission energy W increases with the gas pressure. All things considered, it should be possible to raise the pressure to 100 mm Hg. Even today, for example, pressures of 20 mm Hg have been realized in an N_2(2+) laser. Lasing has been very recently effected in air at atmospheric pressure [123].

The simultaneous improvement of all the factors entering into (63) should produce a significant increase in the energy W (up to about 0.1 J/cm^3 for $\Delta E_{12} = 5$ eV, p = 100 mm Hg, and $\langle\sigma_{01}v\rangle/\langle\sigma_{21}v\rangle = 0.1$).

The problem of increasing the energy of the stimulated emission pulse in pulsed gas lasers is tantamount, by and large, to the problem of elevating the pressure in the pulsed discharge and of creating a homogeneous plasma filling a sufficient volume. As already indicated in Chap. IV, elevation of the pressure entails patent difficulties. The principal difficulty emerges from the fact that the pulsed discharge acquires a strong filamentation tendency when the pressure is increased.

We conclude this section with an examination of one final problem. In the analysis of the N_2(1+) laser operation we generally neglected the radiative decay of the upper working levels. In some situations, however, the radiative decay can prove to be the dominant avenue of de-excitation of the upper laser level. This situation is anticipated for transitions in the short-wave region of the spectrum and, more particularly, for resonance (C–X) transitions in molecular hydrogen (the Werner band system 1100 Å < λ < 1250 Å), in which, as indicated in our paper [27], it is reasonable to expect pulsed lasing (the general pattern of the potential-energy curves for the C and X states and the inversion mechanism are clear from Fig. 35). Appreciable step-by-step filling by electrons via the excitation of a short-lived negative ion [124] should only be expected for the v'' = 1 level. All other molecular levels, judging from the data of [124], should not be as strongly excited.

When radiative decay with probability A is the main channel of de-excitation of the upper laser state, the population equation for the upper (1) and lower (2) laser levels have the form

$$\dot{N}_1 = N_0 n_e \langle\sigma_{01}v\rangle - N_1 A, \tag{64}$$

$$\dot{N}_2 = N_1 A, \tag{65}$$

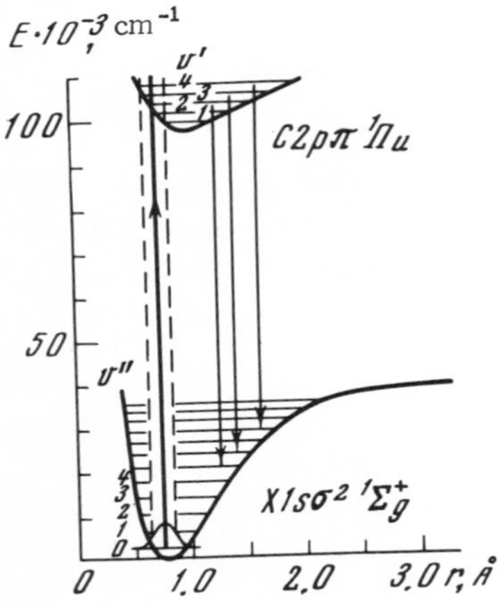

Fig. 35. Potential-energy curves for the X and C states of molecular hydrogen.

where N_0, N_1, and N_2 are the populations of the ground and upper and lower laser levels, respectively (σ_{02} is assumed to be equal to zero). Given an exponential growth of the current and a constant field, for the inversion ΔN we readily obtain

$$\Delta N = N_0 n_e \langle \sigma_{01} v \rangle \frac{\gamma - A}{\gamma(\gamma + A)} e^{\gamma t}, \tag{66}$$

where $\gamma = \alpha v_{dr}$. This equation leads at once to a very significant quantitative result, which previously had only been qualitatively apparent. In order to realize inversion in a transition with radiative decay probability A not only must there be a favorable ratio of cross sections for the upper and lower laser states, but also the growth rate of the electron density per electron ($\dot{n}_e/n_e = \gamma = \alpha v_{dr}$) must exceed the transition probability A. This immediately determines the threshold value of E/p and imposes definite requirements on purely technical parameters such as the actuation speed of the current switch, circuit inductance, etc.

The radiative lifetime of the C state of H_2, computed according to Eq. (4), is 0.5 nsec (see Chapter III). In order to obtain inversion in the C-X transition, according to (66), it is necessary to have $\gamma > 2 \cdot 10^9$ sec^{-1}. The latter growth rate of the electron density is well within the realm of possibility for large enough values of E/p, provided that the voltage pulse has a leading edge of 10^{-10} sec duration. It is clear that the excitation technique described in the present article cannot be used to stimulate laser emission in the Werner bands of H_2. In principle, however, laser emission of this type is, clearly, technically feasible.

§3. Ultimate Efficiency of the $N_2(1+)$ Laser

In determining the laser efficiency it is important to know what fraction of the total energy input to the discharge (per cubic centimeter) is spent for excitation of the upper working state. Within the framework of the pulsed discharge model used here we can readily calculate the total quantity of energy $\mathscr{E}(t)$ developed in 1 cm^3 of the plasma volume at the time t. It must be borne in mind, as shown by our estimates, that the fractional energy spent in the formation and sustainment of the cathode spot (for $t \geq 0.1$ μsec) is negligible by comparison with the energy evolved in the entire plasma volume. The expression for $\mathscr{E}(t)$ may be written in the form

$$\mathscr{E}(t) = \int_0^t e n_e(t) v_{dr} E^{field} dt = \int_0^t \frac{n_e e^2 (E^{field})^2}{m_e N_0 \langle \sigma_t v \rangle} dt. \tag{67}$$

For the case E^{field} = const and the known function $n_e(t)$ Eq. (67) can be used to easily calculate $\mathscr{E}(t)$. Under conditions such that the role of the inductive voltage drop is negligible satisfactory agreement with the experimental is observed over the entire time interval of maximum interest, i.e., the interval of inversion.

Relying on the same discharge model and knowing $\mathscr{E}(t)$, we can readily calculate the excitation efficiency η of the $B_{v'}$ state. Inasmuch as the losses of electron energy in elastic collisions for $\varepsilon_{av} \sim 3$ to 7 eV are considerably lower than in inelastic collisions (by about 10^{-3}), while the diffusion and recombination losses can be neglected, it is reasonable to consider that the total energy E(t) developed per 1 cm^3 is spent in inelastic collisions with molecules, i.e., the following equation must hold:

$$\mathscr{E}(t) = \sum_i \int_0^t n_e N_0 \langle \sigma^i_{in} v \rangle \mathscr{E}^i dt. \tag{68}$$

The summation is taken over all i inelastic collisions of electrons with nitrogen molecules with energy loss $\mathscr{E}^i$. Of all the energy spent in inelastic collisions, only part is directed to excitation of the upper laser B state:

$$\eta_B = \frac{\mathscr{E}_B(t)}{\mathscr{E}(t)} = \frac{\int_0^t N_0 n_e \langle \sigma_{XB} v\rangle \mathscr{E}_B dt}{\int_0^t \frac{n_e e^2 E^2}{m_e} \frac{1}{N_0 \langle \sigma_t v\rangle} dt} \tag{69}$$

or

$$\eta_B = \frac{\int_0^t N_0 n_e \langle \sigma_{XB} v\rangle \mathscr{E}_B dt}{\int_0^t e n_e v_{dr} E dt}. \tag{70}$$

For an exponential growth of the electron density and E^{field} = const the expression for η_B acquires the form

$$\eta_B = \frac{N_0^2}{(E^{field})^2} \frac{m_e}{e^2} \langle \sigma_{XB} v\rangle \langle \sigma_t v\rangle \mathscr{E}_B \tag{71}$$

or, proceeding from (70),

$$\eta_B = \frac{N_0 \langle \sigma_{XB} v\rangle}{e v_{dr} E} \mathscr{E}_B. \tag{72}$$

Equation (72) is better suited to actual calculations. The efficiency η_B calculated according to (72) or its equivalent (71) is less than unity for the physically realizable plasma, because in a real plasma used for the excitation of inversion the energy input per 1 cm^3 is always greater than the energy spent in excitation of the given state. Otherwise, a self-sustained discharge simply could not exist.

The maximum excitation efficiency of all vibrational levels of the B state, calculated according to Eq. (72), attains 35%. The excitation efficiency of the individual vibrational level, of course, is lower. For the B_1 state it is 5%. This figure can be compared with the experimental data. Specifically, knowing the absolute population of the B_1 state and calculating the energy developed in the discharge on the basis of the current pulse, we can calculate the excitation efficiency of the B_1 state. The time variation of the excitation efficiency of this state is shown in Fig. 36 (curves 2 and 3). Up to a time of 1.2 to 1.4 μsec the agreement with the calculation according to Eq. (72) is satisfactory. The subsequent loss of efficiency is attributable to the inception of the B_1-state de-excitation mechanism. For the three strongest laser emission bands, (0, 0), (1, 0), and (2, 1), the total maximum excitation efficiency of the corresponding vibrational levels $B_{0,1,2}$ is about 15%. This figure remains roughly constant up to values of the parameter $E/p \sim 130$ V/cm • mm Hg (Fig. 37, curve 3), then begins to diminish. For $E/p < 60$ V/cm • mm Hg the main fractional energy is spent by electrons in the excitation of ground-state vibrational levels (curve 6). For large values of E/p the losses in excitation of the B state increase (curve 2), while the losses in the vibrational levels decrease. The electron energy losses due to elastic collisions with molecules are smaller than the losses in excitation of the B state by a factor of about $1/5 \cdot 10^2 = 2 \cdot 10^{-3}$.

Knowing the maximum efficiency $\eta_{B_{0,1,2}}$ of excitation of the $B_{0,1,2}$ levels, we can readily calculate the ultimate laser efficiency:

$$\chi_{lim} = \eta \frac{h\nu}{\mathscr{E}_{B_1}} \frac{g_A}{g_B + g_A}, \tag{73}$$

where $h\nu/\mathscr{E}_{B_1} \cong 0.2$, $g_A/(g_B+g_A) = 0.3$. For χ_{lim} we obtain a value $\chi_{lim} \sim 1\%$.

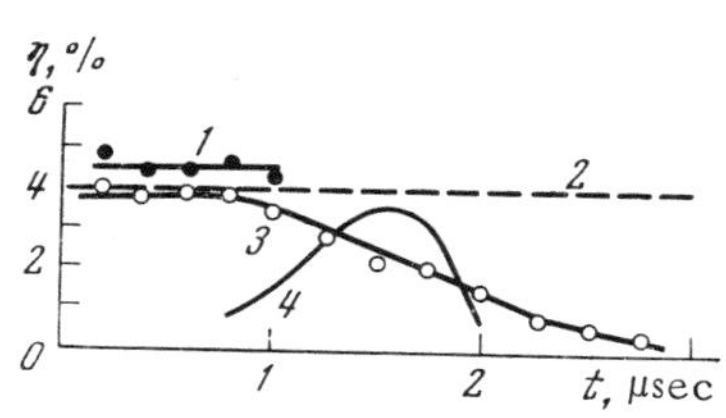

Fig. 36. Excitation efficiency η of B state versus time. 1) Fraction (%) of the energy admitted per 1 cm^3 spent in ionization and subsequent heating of the newly formed electrons; 2) calculated; 3) measured; 4) for comparison, time variation of the population inversion.

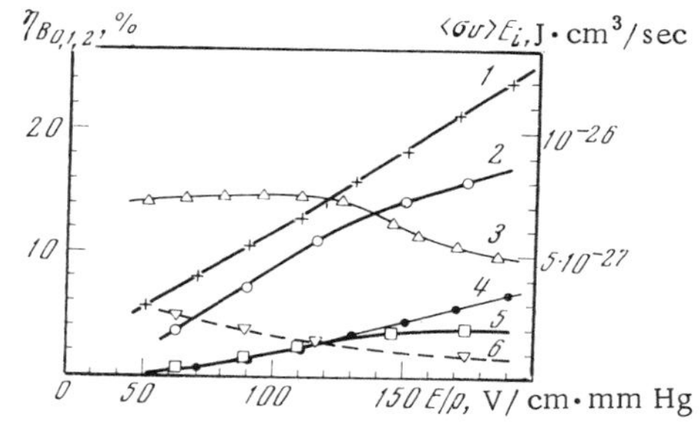

Fig. 37. Excitation efficiency η of $B_{0,1,2}$ levels and electron energy losses in elastic and inelastic collisions versus parameter E/p (calculated). 1) Electron energy losses in elastic collisions with molecules, $(2m/M)\varepsilon_{av}\langle\sigma_{el}v\rangle \cdot 5 \cdot 10^2$; 2) losses in excitation of B state; 3) excitation efficiency of $B_{0,1,2}$ levels; 4) ionization losses; 5) losses in excitation of C state; 6) losses in excitation of ground-state vibrational levels.

It is especially important to stress the fact, as demonstrated above, that up to about 30% of all the energy admitted to the pulsed discharge can be used in the excitation of one electron state (the B state of molecular nitrogen). For the investigated laser a large excitation efficiency of the upper electron state does not, for a great many reasons (relatively low ratio of the radiation quantum to the excitation energy, inefficient operation of a sizable number of bands, etc.), lead to a high overall efficiency. It is entirely conceivable, however, that systems can exist for which the relationship of the indicated factors is more favorable, so that higher overall efficiencies can be realized.

CONCLUSION

We have systematically investigated pulsed laser emission at the leading edge of a powerful current pulse in the molecular gases H_2, D_2, HD, and N_2 (first positive band system of the last). The results strongly support the "direct electron impact" inversion mechanism in these diatomic molecules. This mechanism has been used to explain not only the laser action *per se* at the leading edge of the current pulse, but also to give a qualitatively valid description of the specific attributes of the vibrational–rotational structure of the stimulated emission spectrum both in the isotopic molecules H_2 and D_2 as well as in the 1+ band system of N_2. On the basis of the "direct electron impact" mechanism we have succeeded in carrying out a quantitative analysis of the operation of the $N_2(1+)$ laser. We have calculated with acceptable accuracy the absolute population and excitation rate of the upper working level of the $N_2(1+)$ laser and predicted the exponential population growth of this state with time at the beginning of the high-current discharge stage. Using the stated mechanism we have been able, within the scope of the adopted physical model of pulsed discharge and mechanism of de-excitation of the upper laser state by electrons, to calculate the gain at the individual rotational line as a function of the time and voltage. The agreement between the calculated and measured variables affords conclusive proof in support of the proposed excitation mechanism and the physical model of discharge.

The pulsed discharge model used in the calculations, of course, does not embrace the entire scope of this physical phenomenon. The model and, hence, all the results of the calculations are applicable to a limited sphere of experimentally feasible pulsed discharge states. The proposed model is applicable only so long as the voltage drop across the internal inductance of the discharge circuit and internal resistance of the current source can be neglected and it is certain that cathodic processes do not play a significant part in the current buildup effect. If these conditions are not met, the dynamics of the physical processes in the investigated lasers will be somewhat modified, and the results obtained in the present study can only be used for estimates. We hasten to add, however, that the above-stated restrictions are more of a technical than a physical character. We have been concerned with the fundamental processes associated with pulsed laser action insofar as they govern the physically attainable parameters of the lasers.

The investigations described here provide an educated approach to the solution of such important problems as the continued enhancement of the peak laser power, efficiency, and laser pulse energy obtainable from pulsed gas-discharge lasers. The possibilities for improving these parameters of pulsed gas lasers, all things considered, have scarcely been exhausted. This statement is equally true of the realization of lasing in the short-wave region of the spectrum, in the vacuum ultraviolet in particular.

The initial phase of the present study was conducted under the direction of P. A. Bazhulin. His support and continued interest have contributed immeasurably to the progress of research on pulsed gas lasers.

The author takes this opportunity to thank G. G. Petrash, I. I. Sobel'man, and S. G. Rautian for valuable discussions on various occasions regarding problems germane to the study, as well as M. A. Mazing for making possible the measurements on the DFS-3 photospectrometer equipment.

LITERATURE CITED

1. A. Javan, W. R. Bennett, and D. R. Herriott, Phys. Rev. Lett., 6:106 (1961).
2. G. G. Petrash and I. N. Knyazev, Zh. Éksp. Teor. Fiz., 45:833 (1963).
3. L. E. S. Mathias and J. T. Parker, Appl. Phys. Lett., 3:16 (1963).
4. H. G. Heard, Nature, 200:667 (1963).
5. P. A. Bazhulin, I. N. Knyazev, and G. G. Petrash, Zh. Éksp. Teor. Fiz., 47:1590 (1964).
6. A. A. Vuylsteke, Elements of Maser Theory, Van Nostrand, New York (1960).
7. J. Combrisson, A. Honig, and C. H. Townes, Compt. Rend., 242:2451 (1956).
8. N. G. Basov and A. N. Oraevskii, Zh. Éksp. Teor. Fiz., 44:1742 (1963).
9. L. E. S. Mathias and J. T. Parker, Phys. Lett., 7:194 (1963).
10. P. A. Bazhulin, I. N. Knyazev, and G. G. Petrash, Zh. Éksp. Teor. Fiz., 49:16 (1965).
11. I. N. Knyazev, FIAN Preprint, A-105 [in Russian] (1965).
12. A. Javan, Advances in Quantum Electronics, Columbia Univ. Press (1961).
13. C. Brachet, D. Decomps, G. Durand, L. Heviard-Dubreuilh, H. Lamain, P. Vasseur, and P. Vautier, Compt. Rend., 255:73 (1962).
14. H. A. H. Boot, D. M. Clunie, and R. S. A. Thorn, Nature, 198:773 (1963).
15. W. Bridges, Appl. Phys. Lett., 4:128 (1964).
16. W. R. Bennett, J. W. Kingston, G. N. Mercer, and J. L. Detch, Appl. Phys. Lett., 4:180 (1964).
17. I. N. Knyazev and G. G. Petrash, Zh. Prikl. Spektrosk., 4:560 (1966).
18. I. N. Knyazev, Zh. Prikl. Spektrosk., 5:178 (1966).
19. W. R. Bennett, Jr., Appl. Opt., Suppl. No. 2 on Chemical Lasers (1965).

20. A. L. Bloom, Appl. Opt., 5:1500 (1966).
21. G. G. B. Garrett, Physics of Quantum Electronics, Conf. Proc., McGraw-Hill, New York (1966).
22. W. B. Bridges and A. N. Chester, Appl. Opt., 4:573 (1965).
23. G. G. Petrash, Zh. Prikl. Spektrosk., 4:395 (1966).
24. V. F. Moskalenko, Author's Abstract of Dissertation [in Russian], Ryazan' (1968).
25. J. D. Shipman, Jr., Appl. Phys. Lett., 10:3 (1967).
26. N. N. Sobolev and V. V. Sokovikov, Usp. Fiz. Nauk, 91:425 (1967).
27. N. A. Bazhulin, I. N. Knyazev, and G. G. Petrash, Zh. Éksp. Teor. Fiz., 48:975 (1965).
28. H. S. W. Massey and E. H. S. Burhop, Electronic and Ionic Impact Phenomena, Academic Press, New York (1952).
29. W. T. Walter, N. Solimen, and M. Piltch, IEEE J. Quantum Electronics, QE-2:474 (1966).
30. G. Gould, Appl. Opt., Suppl. No. 2 on Chemical Lasers, p. 59 (1965).
31. V. P. Tychinskii, Usp. Fiz. Nauk, 91:389 (1967).
32. J. Wilson, Appl. Phys. Lett., 8:159 (1966).
33. G. H. Dieke and R. W. Blue, Phys. Rev., 47:261 (1935).
34. I. P. Zapesochnyi and V. V. Skubenich, Opt. i Spektrosk., 21:140 (1966).
35. V. V. Skubenich [Scubenich] and I. P. Zapesochnyi [Zapesochny], Fifth Internat. Conf. Electronic and Atomic Collisions, Abstracts of Papers, Nauka (1967).
36. I. N. Knyazev, FIAN Preprint, A-72 [in Russian] (1965).
37. R. A. McFarlane, Physics of Quantum Electronics, Conf. Proc., McGraw-Hill, New York (1966).
38. K. Bockasten, T. Lundholm, and O. Andrade, J. Opt. Soc. Am., 56:1260 (1966).
39. W. Thompson and S. E. Williams, Proc. Roy. Soc. (London), A147:583 (1934).
40. H. G. Cooper and P. K. Cheo, Appl. Phys. Lett., 5:42 (1964).
41. T. Kasuya and D. R. Lide, Jr., Appl. Opt., 6:69 (1967).
42. D. A. Leonard, Appl. Phys. Lett., 7:4 (1965).
43. V. M. Kaslin and G. G. Petrash, ZhÉTF Pis. Red., 3:88 (1966).
44. E. T. Gerry, Appl. Phys. Lett., 7:6 (1965).
45. H. G. Cooper and P. K. Cheo, Appl. Phys. Lett., 5:42 (1964).
46. A. Henry, G. Arya, and L. Henry, Compt. Rend., 261:1495 (1965).
47. M. Huber, Phys. Lett., 12:102 (1964).
48. R. A. McFarlane, Phys. Rev., 140:A1070 (1965).
49. L. B. Loeb, Fundamental Processes of Electrical Discharge in Gases, Wiley, New York (1939).
50. V. M. Sutovskii, Author's Abstract of Candidate's Dissertation [in Russian], FIAN SSSR (1968).
51. A. S. Nasibov, A. A. Isaev, V. M. Kaslin, and G. G. Petrash, Pribory i Tekh. Éksperim., 4:232 (1967).
52. W. F. Meggers, J. Res. Nat. Bur. Standards, 10:427 (1933).
53. A. H. Poetker, Phys. Rev., 30:418 (1928).
54. G. H. Dieke, Phys. Rev., 50:797 (1936).
55. G. H. Dieke and J. Molek, Spectroscopy, 2:494 (1958).
56. G. H. Dieke and D. F. Heath, Johns Hopkins Spectroscopic Rep. No. 17, Baltimore, Md. (1959).
57. P. K. Carrol, Proc. Roy. Irish Acad., 54A:369 (1954).
58. J. C. DeVos, Physica, 20:690 (1954).
59. H. R. Briem, Plasma Spectroscopy, McGraw-Hill, New York-London (1964).
60. I. I. Sobel'man, Introduction to the Theory of Atomic Spectra, Fizmatgiz (1963).
61. E. A. McLean and S. A. Ramsden, Phys. Rev., 140:A1122 (1965).
62. J. B. Gerardo and J. T. Verdeyen, Proc. IEEE, 52:690 (1964).

63. D. S. Rozhdestvenskii, Papers on Anomalous Dispersion in Metal Vapors, Izd. AN SSSR (1951).
64. G. H. Gale, G. S. Monk, and L. O. Lee, Astrophys. J., 67:89 (1928).
65. G. Herzberg, Molecular Spectra and Molecular Structure; Spectra of Diatomic Molecules (2nd ed.), Van Nostrand, New York (1961).
66. J. Tobias and J. T. Vanderslice, J. Chem. Phys., 35:1852 (1961).
67. E. R. Davidson, J. Chem. Phys., 35:1189 (1961).
68. G. H. Dieke and R. W. Blue, Phys. Rev., 76:50 (1949).
69. R. W. Nicholls, J. Quant. Spectrosc. Radiative Transfer, 2:433 (1962).
70. V. N. Soshnikov, Usp. Fiz. Nauk, 74:61 (1961).
71. R. Milliken and C. Rieke, Rep. Progr. Phys., 8:231 (1941).
72. R. Roscoe, Phil. Mag., 31:349 (1941).
73. J. D. Craggs and H. S. W. Massey, Handbuch der Physik, 37/1, Springer, Berlin (1959). p. 334.
74. L. V. Leskov and F. A. Savin, Usp. Fiz. Nauk, 72:741 (1960).
75. D. D. Lambert, Atomic and Molecular Processes [Russian translation], Izd. Mir (1964).
76. I. N. Knyazev, Zh. Prikl. Spektrosk., 3:510 (1965).
77. C. K. N. Patel, Phys. Rev. Lett., 12:588 (1964).
78. W. R. Bennett, Appl. Opt., Suppl. No. 1 on Optical Masers, pp. 24-62 (1962).
79. É. M. Belenov and A. N. Oraevskii, Opt. i Spektrosk., 18:858 (1965).
80. M. Haude, Proc. Roy. Soc., 136:114 (1932).
81. O. Andrade, M. Gellardo, and K. Bockasten, Appl. Opt., 6:2006 (1967).
82. J. T. Vanderslice, E. A. Mason, and E. R. Lippincott, J. Chem. Phys., 30:129 (1959).
83. W. Lichten, J. Chem. Phys., 26:306 (1957).
84. N. P. Carleton and O. Oldenberg, J. Chem. Phys., 36:3460 (1962).
85. W. H. Wurster, J. Chem. Phys., 36:2111 (1962).
86. A. P. Dronov, N. N. Sobolev, and F. S. Faizullov, Opt. i Spektrosk., 21:538 (1966).
87. R. W. Nicholls, J. Res. Nat. Bur. Standards, 65A:451 (1961).
88. M. Jeunehomme and A. B. F. Duncan, J. Chem. Phys., 41:1692 (1964).
89. A. Budo, Z. Phys., 105:579 (1937).
90. A. Budo, Z. Phys., 96:219 (1935).
91. R. Seeliger and K. Bock, Z. Phys., 110:717 (1938).
92. L. N. Tunitskii and A. I. Ignashkov, Svetotekhnika, No. 2, p. 23 (1955).
93. A. V. Nedospasov and A. E. Novik, Zh. Tekh. Fiz., 30:1329 (1960).
94. F. H. Mitchell and L. B. Snoddy, Phys. Rev., 72:1232 (1947).
95. R. A. Westberg, Phys. Rev., 114:1 (1959).
96. G. V. Spivak and E. L. Stolyarova, Zh. Tekh. Fiz., 20:501 (1950).
97. É. M. Reikhrudel', A. V. Kustova, and A. G. Zimelёv, Zh. Tekh. Fiz., 24:1179 (1954).
98. A. E. Novik, Svetotekhnika, No. 12, p. 4 (1962).
99. F. Llewellyn Jones, Handbuch der Physik, 22/2, Springer, Berlin (1956).
100. J. M. Meek and J. D. Craggs, Electrical Breakdown of Gases, Clarendon Press, Oxford (1953).
101. I. N. Knyazev, Élektron. Tekh., Ser. 3, Gazorazryadnye Pribory, No. 2, p. 3 (1967).
102. B. N. Klyarfel'd, A. G. Guseva, and A. S. Pokrovskaya, Zh. Tekh. Fiz., 36:704 (1966).
103. C. W. McClure, Phys. Rev., 124:969 (1961).
104. C. W. Paxton and R. G. Fowler, Phys. Rev., 128:993 (1962).
105. M. A. Sultanov and L. I. Kiselevskii, Zh. Prikl. Spektrosk., 1:268 (1964).
106. L. B. Loeb, Phys. Rev., 113:7 (1959).
107. V. L. Ginzburg and A. V. Gurevich, Usp. Fiz. Nauk, 70:201 (1960).
108. L. A. Artsimovich, Controlled Thermonuclear Reactions, Fizmatgiz (1961).
109. A. G. Engelhardt, A. V. Phelps, and C. G. Risk, Phys. Rev., 135:A1566 (1964).

110. J. Hasted, Physics of Atom Collisions, Butterworths, London (1964).
111. L. Frommhold, Z. Phys., 160:554 (1960).
112. A. E. D. Heylen, Proc. Phys. Soc., 79:508 (1962).
113. M. A. Harrison, Phys. Rev., 105:366 (1957).
114. E. W. McDaniel, Collision Phenomena in Ionized Gases, Wiley, New York (1964).
115. J. S. Townsend and V. A. Bailey, Phil. Mag., 42:873 (1921).
116. Yu. M. Kagan and R. I. Lyagushchenko, Zh. Tekh. Fiz., 31:445 (1961).
117. A. V. Gurevich, Zh. Éksp. Teor. Fiz., 37:304 (1959).
118. Yu. M. Kagan and R. I. Lyagushchenko, Zh. Tekh. Fiz., 32:735 (1962).
119. A. Engel, Handbuch der Physik, 21/1, Springer, Berlin (1962), p. 504.
120. D. T. Stewart, Proc. Phys. Soc., A68:404 (1955).
121. A. W. Ali, A. C. Colb, and A. D. Anderson, Appl. Opt., 6:2115 (1967).
122. H. Regemorter, Astrophys. J., 136:906 (1962).
123. A. Svedberg, L. Höberg, and R. Nilsson, Appl. Phys. Lett., 12:102 (1968).
124. G. J. Shulz, Phys. Rev., A135:988 (1964).